POWER SYSTEM OPERATIONS and ELECTRICITY MARKETS

The ELECTRIC POWER ENGINEERING Series
Series Editor Leo Grigsby

Published Titles

Electromechanical Systems, Electric Machines, and Applied Mechatronics
Sergey E. Lyshevski

Electrical Energy Systems
Mohamed E. El-Hawary

Electric Drives
Ion Boldea and Syed Nasar

Distribution System Modeling and Analysis
William H. Kersting

Linear Synchronous Motors: Transportation and Automation Systems
Jacek Gieras and Jerry Piech

The Induction Machine Handbook
Ion Boldea and Syed Nasar

Power Quality
C. Sankaran

Power System Operations and Electricity Markets
Fred I. Denny and David E. Dismukes

Forthcoming Titles

Computational Methods for Electric Power Systems
Mariesa Crow

POWER SYSTEM OPERATIONS and ELECTRICITY MARKETS

FRED I. DENNY
DAVID E. DISMUKES

CRC PRESS

Boca Raton London New York Washington, D.C.

Library of Congress Cataloging-in-Publication Data

Denny, Fred I.
Power system operations and electricity markets / Fred I. Denny, David E. Dismukes.
p. cm.
Includes bibliographical references and index.
ISBN 0-8493-0813-5
1. Electric power production. 2. Electric utilities—United States. 3. Electric power—United States—Marketing. I. Dismukes, David E. II. Title.

TK1001 .D45 2002
333.793'2'0973—dc21 2002276807
CIP

Direct all inquiries to CRC Press LLC, 2000 N.W. Corporate Blvd., Boca Raton, Florida 33431.

Visit the CRC Press Web site at www.crcpress.com

International Standard Book Number 0-8493-0813-5
Library of Congress Card Number 2002276807
Printed in the United States of America 1 2 3 4 5 6 7 8 9 0
Printed on acid-free paper

Preface

Prior to the 1970s, the U.S. electric power industry was technology driven. Engineers were trained to focus on specific technologies and work in specialized areas. However, dramatic changes began taking place in the 1970s, and an "energy crisis" ushered in a new era of tighter regulation.

By the early 1990s, two decades of intense regulation were replaced by a new policy of promoting open access and competition. The Energy Policy Act of 1992, followed by several significant Notices of Proposed Rulemakings and Orders from the U.S. Federal Energy Commission, required utilities to compete for markets that were previously guaranteed. As a consequence, there were many mergers and acquisitions. Marketers with little or no power industry experience moved into positions of influence. Engineering organizations were downsized, and the engineers who were left behind had to find ways to prevent power systems from becoming less reliable.

Today the transition in industry structure is nearly complete. The U.S. electric power industry is no longer technology driven. It is public policy and market driven. Just as utility companies have to change to survive in the new competitive environment, engineers who choose to work in the industry must learn a new set of skills and accommodate new disciplines.

This book is intended to help educate new engineers for the new business environment. We explain how new methods for power systems operations and energy marketing relate to public policy, regulation, economics, and engineering science. This book can serve as a textbook for an undergraduate elective course for engineering students. Alternatively, it can be used for the continuing education of industrial power engineers and energy industry employees.

About the Authors

Dr. Fred I. Denny is currently associate professor of electrical engineering at McNeese State University, Lake Charles, LA. He took a leave of absence from the university in the spring and summer of 2001 to work for the North American Electric Reliability Council (NERC) in Princeton, NJ. From 1995 to 2000, he served as an associate professor of electrical engineering at Louisiana State University in Baton Rouge. From 1979 to 1995, he was vice president of engineering for Edison Electric Institute (EEI) in Washington, D.C. Before joining EEI, he was manager of the operating services support department at Southern Company Services in Birmingham, AL. He also served as a captain in the United States Army Reserve. Dr. Denny is an IEEE Fellow and a licensed professional engineer in Louisiana, Alabama, and Georgia. He is a member of the National Society of Professional Engineers, the Louisiana Engineering Society, and the IEEE Power Engineering Education Committee. He serves on the editorial board for IEEE computer applications in *Power Magazine* and is secretary of the CIRED U.S. National Committee. Dr. Denny earned a Ph.D. in electrical engineering from Mississippi State University, specializing in power systems engineering. He is the author of many publications and articles on power systems engineering issues and technologies.

Dr. David E. Dismukes is currently associate professor at the Center for Energy Studies (CES), Louisiana State University (LSU). His primary research interests are related to policy issues in regulated and energy industries. He serves as a policy adviser to government agencies and private industry, and speaks regularly before professional and civic associations on electric restructuring issues. He has published many articles on energy and regulated industries issues. Prior to joining the CES faculty in 1995, Dr. Dismukes served as a staff economist at the Florida Public Service Commission and as a research associate with a nationally recognized economic consulting firm located in Tallahassee, FL. Dr. Dismukes is a member of Omicron Delta Epsilon, the American Economic Association, the American Statistical Association, The Econometric Society, The Southern Economic Association, the Western Economic Association, and the International Association of Energy Economists. In addition to his CES faculty appointment, Dr. Dismukes is a member of the LSU graduate faculty and is an adjunct associate professor in the department of economics in the E.J. Ourso College of Business Administration at LSU. He earned his B.A. from the University of West Florida in Pensacola, and his M.S. and Ph.D. degrees (economics) from Florida State University in Tallahassee.

Contents

chapter one

The evolution of the electric power industry

1.1 Energy conservation in the pre-energy crises environment

During the past 75 years, the experiences of the electric power industry have been heavily conditioned by economic regulation at the state and federal levels. Starting in the 1920s, policymakers at the state level* began subjecting power utilities within their respective jurisdictions to regulatory oversight based upon two premises: (1) the industry had natural monopoly cost characteristics, and (2) the industry was imbued with the public interest. These premises supported the continued regulation of power markets through the 1990s when a number of these underlying premises, particularly the notion of the industry being characterized as a natural monopoly, began to unravel.

A natural monopoly is a special case in the economic organization of markets. A natural monopoly is perhaps the only case where allowing one firm to operate is more efficient than promoting production between several firms. When a natural monopoly exists, the technology of production is such that economies of scale (declining average cost per unit of output) are said to exist over the entire range of market demand for that good or service. If this firm was broken into several smaller firms, these economies (and lower average costs) would not be achieved, and prices to end users would be higher than if the good or service was produced by only one company. Hence, early support for regulation was built on trying to maintain this natural monopoly, and at the same time tempering its potential excesses.

The problem with natural monopolies is that if left unchecked, they have the ability to increase costs to levels that are considerably higher than current costs. Figure 1.1 presents the cost and pricing characteristics of natural monopoly firms. Prices and costs are represented on the vertical axis, while

* There is a corresponding level of regulation at the federal level that began in the 1920s with the passage of the Federal Power Act (FPA) and its subsequent revisions in the 1930s, in addition to the passage of the Public Utilities Holding Companies Act (PUHCA).

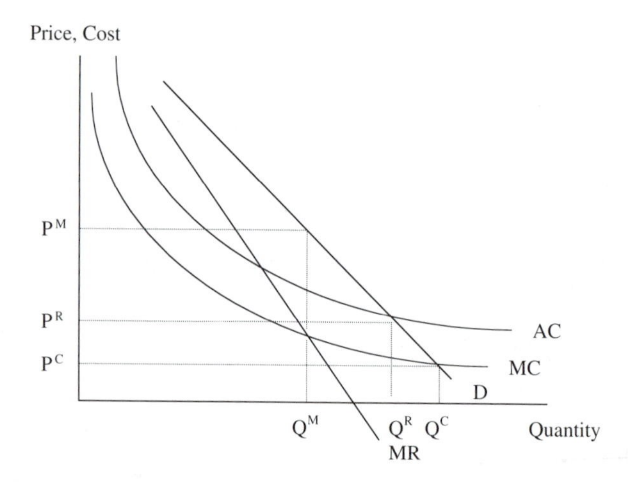

Figure 1.1 Natural monopoly cost and pricing characteristics.

quantity is represented on the horizontal axis. The natural monopoly characteristics show that average costs (AC) and marginal costs (MC) are decreasing throughout the entire range of production for the natural monopoly firm. The downward-sloping line marked by D is the demand curve. In perfectly competitive markets, firms will set prices equal to costs — in this case, marginal costs. In equilibrium, prices will be set at P^C, and quantity supplied/demanded will be set at Q^C.

To maximize profits, natural monopolies will typically set prices higher than what should occur under competitive outcomes. Below the demand curve, represented by D, is the marginal revenue curve, represented by MR. This curve maps the incremental revenues associated with additional sales over a given price. Profit-maximizing monopolies will price their services at the point where MR intersects MC on the demand curve. This leads to an alternative price and quantity combination represented by P^M and Q^M, respectively. Here, prices are higher and quantity supplied is lower than under competitive market conditions. It is this type of potential abuse that regulation is designed to forestall.

The problem with regulation is that it cannot perfectly match competitive market conditions. As shown by Figure 1.1, pricing at MC = D will result in a loss of cost recovery in a number of fixed costs (i.e., the difference between AC and MC). As a result, regulators will set prices equal to average costs. This results in slightly higher prices than under competitive market conditions, but will allow the natural monopoly to remain in business and tempers its potential profit-maximizing pricing levels.

During the turn of the century, industrializing nations recognized that there were two primary means of regulating large natural monopoly industries like electricity, telecommunications, natural gas, water, and wastewater. The first means was through nationalization. Under such an approach, the industry in question would be owned and operated by the national, or sometimes regional, government institution. Such an approach was followed in many countries in Europe and Latin America. Nationalization is premised upon the belief that the government will be able to operate a natural monopoly effectively and in the public interest. The common criticism of this form of regulation, however, is that public institutions do not face profit-maximizing incentives to keep costs down. In later decades, this criticism was seen by many as particularly true and led to calls for privatization.

For the most part, the United States' model for containing potential natural monopoly abuses has rested with price and earning regulations of private industries. This approach is commonly referred to as rate of return (ROR) regulation. Under the ROR model, utilities are allowed to set prices in a manner that allows them to recover their ongoing operational costs, as well as the opportunity to earn a reasonable ROR on their investments. Prices are set on an average cost basis that includes both of these cost components. This method worked well through the better part of the 20th century, particularly in the electric power industry. During this period, the industry was able to garner significant economies of scale in the production, transmission, and distribution of electricity. A number of benefits to utilities were associated with pushing technological innovations. If utilities could lower costs while keeping rates constant, then they could increase profits between periods where no regulatory rate cases existed and maximize earnings for their shareholders. It has been noted that during this period regulators tended to pursue a live-and-let-live policy with regard to utilities. The primary concern of regulators was to keep nominal prices from increasing. Firms were allowed to earn generous rates of return if they could increase their achievement rates without raising prices, provided that costs continued to decrease for their captive ratepayers.[1]

ROR regulation is not without its own set of criticisms, many of which became strikingly evident in the late 1970s and early 1980s. The primary criticism levied against traditional or ROR regulation rests with the "overcapitalization" hypothesis. As regulated utilities, these firms have incentives to make significant investments to increase their overall earnings. The higher the investment, the higher the overall allowed returns. Such a regulatory approach could lead to "gold plating" and overinvestment.

The idea of gold plating, or overcapitalization, was the attention of much scholarly debate during the 1960s and 1970s. Averch and Johnson (1962) formed a static, deterministic model of the regulated firm subject to a regulatory constraint. The regulatory constraint is merely a cap, set by the regulatory body, on the maximum allowable ROR that the regulated firm can earn. In the model, depreciation is assumed to be zero, and the only cost of acquiring capital is the interest to be paid on the plant and equipment.

After formulating this model, Averch and Johnson reached two controversial conclusions. Specifically, they concluded that a regulatory bias exists, which encourages the regulated firm to make inefficient capital-intensive investments.[2] A second, but often overlooked conclusion is that regulated firms also have the incentive to cross-subsidize less profitable operations at the expense of its more profitable operations, as long as the firm's overall rate of return remains unchanged. Both of these conclusions provide formal evidence that ROR regulation can impose social costs in the form of input and output inefficiencies.

The debate on whether ROR regulation imposed a negative incentive on regulated firms and their investment and operational strategies remained relatively academic until the 1970s. Most electric utilities prior to the 1970s were in a decreasing cost environment, and their rates reflected these productivity advantages. The energy crises of the 1970s and early 1980s, however, shifted this debate to the front burner — not only for academics, but for policymakers as well.

1.2 *The energy crisis and its impact on the electric power industry*

After 1973, the electric power industry entered a much more volatile and less predictable period. The oil embargo by the Organization of Petroleum Exporting Countries (OPEC) increased primary energy prices to utilities and, subsequently, to their retail customers. Rate increases, reflecting large increases in costs, significantly dampened the growth in electricity demand. Technological advances slowed, and the only new steam technology being developed at the time, nuclear power, was reeling from a number of accident-related setbacks, including the fire at the Tennessee Valley Authority's (TVA) Browns Ferry unit (1975) and the infamous accident at the Three Mile Island plant in Pennsylvania (1979). Regulatory policy at both the state and federal levels shifted from a relatively passive oversight role to a more micromanaged, adversarial mode. In addition, policy took a major shift in 1978 with the enactment of the Public Utilities Regulatory Policy Act (PURPA). This policy, which was originally intended to stimulate greater on-site energy efficiency at industrial facilities, sowed the seeds for competition, which began to blossom over the next two decades.

1.2.1 *Economic factors influencing the electric power industry*

The most significant economic changes after 1973 included the costs of electric generation, fuel prices, the demand for electricity, the capital costs of constructing electric generation facilities, and the costs of financing large construction projects. All of these factors played a significant role in undermining the natural monopoly cost characteristics of the industry and created opportunities for both energy conservation and competition.

If the real retail rate of electricity is used as an approximation of the cost of electric generation,* it is easy to see two distinct historic trends in utility costs. Figure 1.2 shows that throughout the 1960s, the average (real) retail

* State regulatory commissions typically set rates at the cost of generation, plus some allowed rate of return on investment.

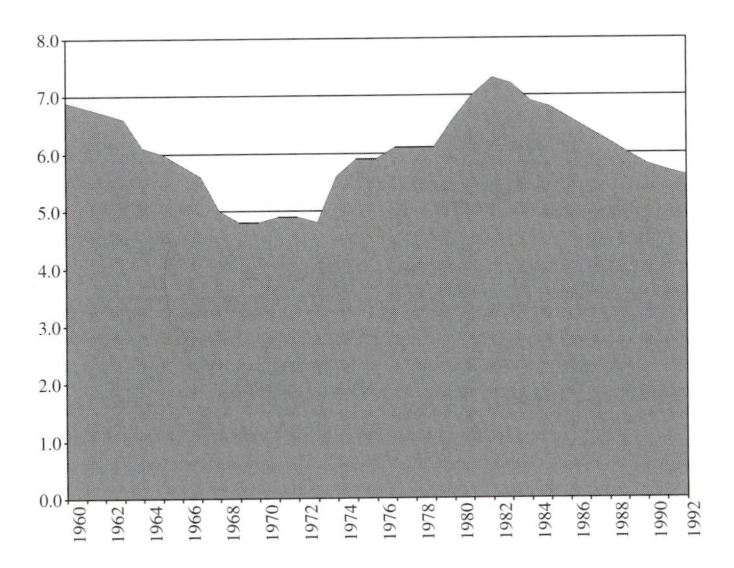

Figure 1.2 Historic electricity prices (1960–1992).

rate of electricity followed a steady and gradual decline. This decline was the result of a number of factors, including low fuel costs and tapping the economies of scale present in large central station electric generation, as well as high voltage transmission. It was the advantages created during this period that allowed utilities to continue to maintain strong earnings for their shareholders, while at the same time keeping regulators happy with decreasing rates.

However, the good times in the power business were soon to end. Between 1973 and1974, the first year following the OPEC oil embargo, electricity prices jumped 17%. The cost of electric generation continued to escalate throughout the late 1970s, mainly because of the increased capital cost of generation. By 1982, the year following the Iranian revolution and a second world price increase, a combination of high capital and fuel costs forced the real price of electricity to an all-time high of 7.3 cents kWh, 52% higher than the pre-1973 rate. Fortunately, these levels were not sustained, and beginning in 1982 the real price of electricity began to fall as fuel prices eased and utility construction programs were all but phased out.* Unfortunately for utilities, these decreases in rates came a little too late for the irreversible changes in market structure that arose in the early to mid-1980s.

One of the primary culprits for increased power generation was associated with fossil fuel prices. Figure 1.3 shows how dramatically real fossil fuel prices (oil, natural gas, and coal) increased in 1973. In 1973, real oil prices were 50% higher than their 1972 levels. On the other hand, coal was only

* Fuel expenses were 67, 84, and 77% of power production expenses (excluding capital expenditures) for major investor-owned utilities (IOUs) in 1990, 1980, and 1970, respectively. Energy Information Administration. Financial Statistics of Selected Investor-Owned Electric Utilities 1990. (Washington: U.S. Department of Energy, 1992): 26. See predecessor issues for 1980 and 1970 expenses.

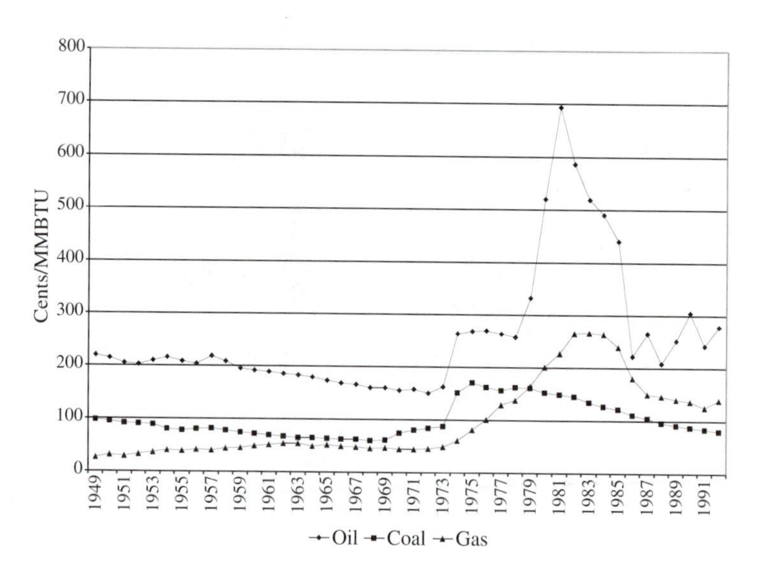

Figure 1.3 Historic fossil fuel prices (cents per MMBTU).

4% higher and natural gas only 25% higher than their respective levels of a year earlier.

Perhaps the most significant post-1973 change for the electric utility industry was the increased capital cost required to construct electric generation facilities. The inflationary environment of the 1970s was a difficult period for all major construction projects, power plants notwithstanding. Figure 1.4 shows the dramatic changes in average construction costs by fuel type, for large baseload steam generation facilities.* In 1981, oil and natural gas ceased to be steam generation options because of the implementation of the Power Plant and Industrial Fuels Act (Fuel Use Act).** In addition, the capital costs associated with the remaining two baseload steam options, coal and nuclear, began to increase significantly after the enactment of the Fuel Use Act. After 1981, coal capital costs increased on average by about 13% per year. Nuclear costs, on the other hand, had annual average increases of about 46%. The relative difference between the two fuels is even more dramatic. For instance, in 1981, the costs of nuclear and coal generation were $746.10/kW and $598.20/kW, respectively. By 1990, the installed cost for a coal generation facility was $1616.70/kW, while the cost for nuclear generation was $4167.90/kW — or 2.5 times as much.

While construction costs increased dramatically during the period, financing costs for these facilities were equally ominous. These increased financing costs, also the result of the stagflationary environment of the 1970s, contributed to escalated power plant capital costs. These costs, which were

* These capital costs are in mixed (inflated) dollars and include interest costs, which are commonly referred to as the "allowance for funds used during construction" (AFUDC).

** While oil and gas are prohibited from use in steam generation, both fuels continue to be used in combustion turbine and combined-cycle generators.

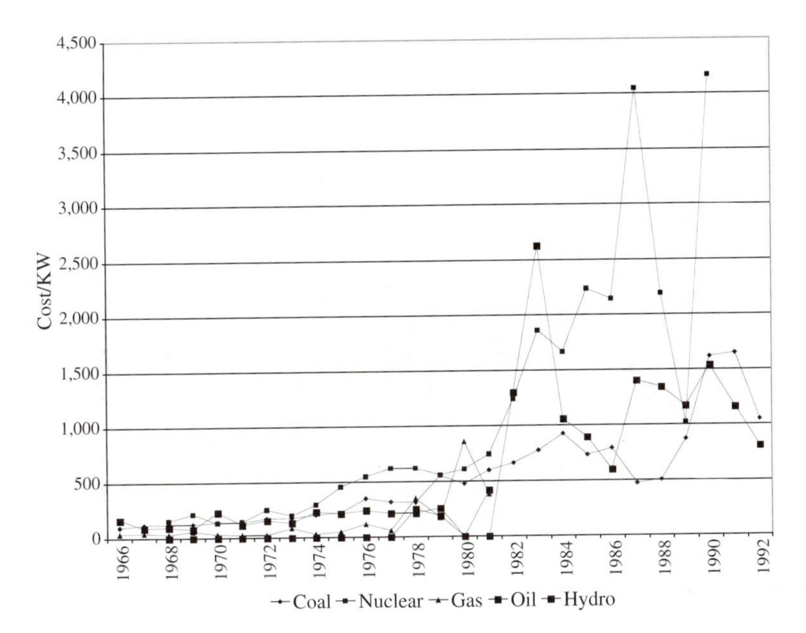

Figure 1.4 Historic installed cost of electric generation facilities (cost/kW). (From Electric Utility Power Plant Construction Costs, Utility Data Institute, Washington, D.C., 1994.)

incorporated into utility rates while construction proceeded, increased with the utility's cost of capital and the lengthening of power plant construction duration. In many instances, financing costs represented as much as 15 to 20% of the real capital investment of a new baseload power plant.[3] Figure 1.5 presents an approximation to these financing costs for the past 20 years, as represented by the weighted average industry cost of capital.

One of the more dramatic results of the 1970s was the significant decrease in consumer energy demand that resulted from the energy crises. This was true across all energy services — electricity included. This came as a shock to the industry, as well as government agencies that regularly forecasted and examined energy usage trends. Most clearly underestimated the price sensitivity that customers have for their electricity services. In the late 1960s, the annual average rate of growth for electricity was between 6 and 8% per year. These trends were seen as increasing into perpetuity and were one of the justifications for the significant nuclear power plant building campaigns upon which many electric utilities embarked.

However, as shown in Figure 1.6, electricity demand did not keep up with its historic trends. The first shock to annual growth of electricity usage began suddenly in 1973. While there was a small rebound from 1976 to 1981, these annual averages were clearly lower than those of the earlier decade. The recession of 1981 to 1982 further facilitated these decreases as the economy contracted, industrial and manufacturing output decreased, and the demand for electricity fell. By 1992, the annual average growth in electricity usage hovered around 2%. This created an environment of excess capacity.

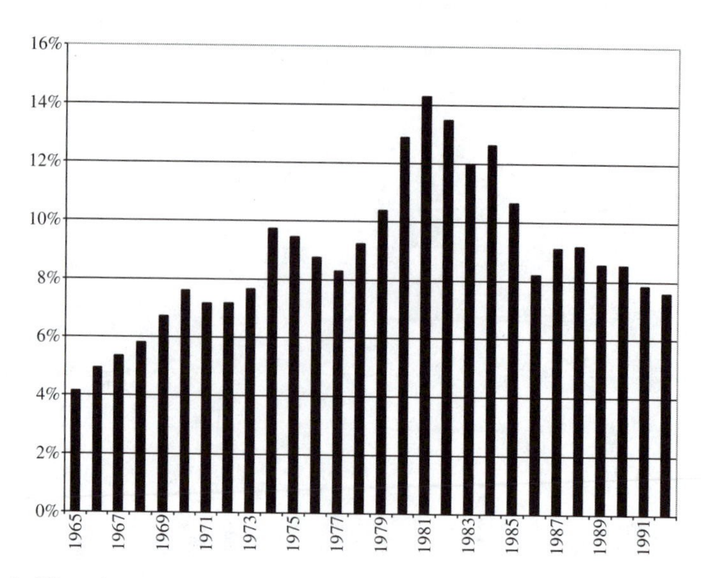

Figure 1.5 Historic average weighted cost of capital.

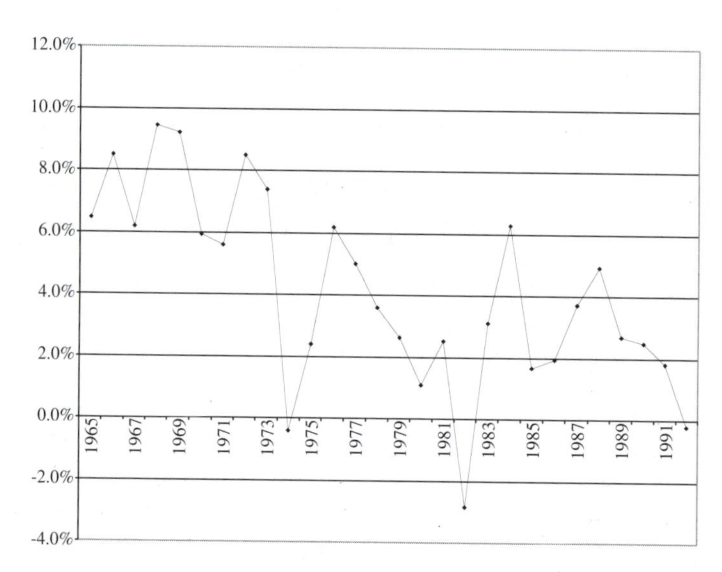

Figure 1.6 Annual growth rates of electricity demand.

Many electric utilities in the 1980s found that they had enormous generating capacity — mostly nuclear — but an anemic customer base upon which to make these sales.

1.2.2 Technological factors influencing the electric power industry

Throughout most of the 1960s, the electric power industry was one of the leading sectors of the economy in terms of technical innovation and productivity growth.[4] The amount of heat input, measured in British thermal units (BTUs), needed to generate a kWh of electricity with steam turbines decreased by almost 40% between 1925 and 1945, and by 35% during 1945 and 1965.[4] Increasing thermal efficiencies enabled utilities to reduce costs. As these developments began to taper after 1965, so did the industry's ability to offset the exogenous economic changes in costs discussed earlier.

Advances in thermal efficiency dramatically enhanced economies of scale in electric generation. However, the last major thermal development, made in the 1960s, was commercially exploited during most of the 1970s.[5] This technology involved the development of the supercritical boiler (used in fossil fuel generation), which could achieve boiler pressures of above 3200 pounds per square inch (psi).

The development of supercritical units represented a dramatic departure from earlier subcritical technologies. Water heated to a temperature of above 706 degrees at a pressure of more than 3200 psi directly vaporizes to dry, superheated steam.[6] This eliminates equipment required to extract saturated steam, recycling equipment, and some equipment to heat saturated steam. Supercritical boilers, however, do require additional expenditures on materials to accommodate the tremendous increase in steam pressure.[6]

From 1970 to 1974, the supercritical boiler achieved a 63% market penetration rate in new installations.[7] This market share started to decline throughout the 1970s, and by 1982, supercritical boilers were being installed in only 6% of new generators.[7] Unanticipated maintenance problems associated with higher pressure units contributed to the decline in the use of supercritical technology. In addition, large decreases in electricity demand, which began in 1973, discouraged many utilities from building ahead of demand with larger, higher-efficiency units.[8]

The only remaining technological development in the electric power industry during this time period was nuclear power. In the 1960s, many utilities began constructing their own nuclear power plants. Table 1.1 shows that prior to 1972, 135 nuclear power plants were ordered for a combined total of 10,829 gigawatts (GW). By 1973, the industry had peaked with 41 orders for new nuclear power units. By 1975, however, canceled units outpaced new orders, and the industry began its long slide toward fewer numbers of orders and larger numbers of cancellations.

Several factors contributed to the demise of nuclear power as a technological option for electric utilities. The most obvious reason rests with the construction cost disadvantages, outlined in Figure 1.4, that nuclear had relative to other competing fuels. Another disadvantage included the lengthy construction duration required to complete a nuclear unit. In 1968, it took 3.3 years to construct a nuclear power plant. By 1973, construction duration had increased to 5.6 years. After 1973, a significant and growing number of safety-related construction requirements pushed construction duration even higher. By 1987, construction duration had reached an all-time high of 11 years.

Table 1.1 Nuclear Power Plant Orders, Cancellations, and Commitments

Year	Orders Placed		Cancellations		Total Commitments	
	Number	GW	Number	GW	Number	GW
1972	38	4153	6	574	160	14,382
1973	41	4683	0	0	201	19,065
1974	26	3093	8	829	218	21,325
1975	4	418	11	1229	211	20,514
1976	3	379	2	233	212	20,656
1977	4	504	9	986	209	20,178
1978	2	224	13	1333	196	19,068
1979	0	0	8	948	188	18,121
1980	0	0	16	1809	171	16,286
1981	0	0	6	581	165	15,705
1982	0	0	18	2202	146	13,497
1983	0	0	6	374	139	12,866
1984	0	0	8	904	130	11,963
1985	0	0	0	0	130	11,952
1986	0	0	3	238	127	11,714
1987	0	0	0	0	126	11,699

By the mid 1980s, construction in most large steam-generation facilities had come to a halt. During this period, a number of technological advances arose in combustion turbine and combined-cycle generation technologies. One could argue that the development and increased use of natural-gas-fired, combined-cycle generation were a result of the high capital cost and regulatory environment of the late 1970s and early 1980s. Combined-cycle plants* represent a significant alternative to traditional baseload generation. Combined-cycle plants can generate a significant amount of capacity and, as such, are not easily categorized as being either baseload, intermediate, or peaking load units.

The popularity of combined-cycle units has increased dramatically over the past several years in both the utility and nonutility generation of electricity. For instance, during the period from 1966 to 1993, combined-cycle plants comprised some 6% of all new power plants, compared to 84% for steam.[9] Today, high-efficiency combustion turbines and combined-cycle units are the technology of choice. Clearly, the significant cost increases for traditional steam generation have resulted in a greater preference for these more modular, more quickly constructed, and environmentally cleaner units.

1.2.3 *Public policy factors influencing the electric power industry*

Certainly, public policy initiatives play a significant role in the decisions made by major industries in the United States. The electric power industry

* Combined-cycle plants are comprised of a gas-fired combustion turbine with a waste heat recycling unit; thus, a combined cycle of electric generation. The first stage generates gas-fired electricity from a turbine, while the second stage captures the waste heat to create steam to drive a second turbine.

is no different. Public policy, however, plays an even greater role in this industry, given its franchised monopoly position and its subjection to economic, safety, and environmental regulation at both the federal and state levels. Dramatic changes in public policies affecting the electric utility industry have occurred at both the state and federal levels of government. A review of such changes would take volumes to cover. The emphasis here is not to document every policy initiative that had an impact on the industry, but rather to focus on the major changes and outline how they influenced the industry.

1.2.3.1 Federal public policy initiatives

Major federal public policy initiatives affecting electric utilities fall within three areas: safety regulation, environmental regulation, and economic regulation, including promoting competition in electric generation. Many of these initiatives were warranted and have resulted in benefits that have exceeded costs. But, for the established electric power industry, many of these initiatives have also increased the cost and precariousness of constructing, as well as operating and maintaining, large electric generation facilities.

Federal safety regulation is an area that dramatically affected the development of nuclear power throughout the 1970s and 1980s. During this period, the Nuclear Regulatory Commission (NRC) issued a significant number of safety-related regulations as a result of accidents at nuclear plants around the country. In 1975, the Tennessee Valley Authority (TVA) Browns Ferry plant caught fire in Alabama. This accident resulted in dramatically different standards for electric separation and fire prevention. In 1979, the infamous near disaster at Three Mile Island (TMI) in Pennsylvania sent regulatory shock waves throughout the industry.

Federal environmental regulations also had dramatic impacts on the electric utility industry. Starting in the early 1970s, the Environmental Protection Agency (EPA) established and strengthened a number of air quality standards that required new coal-fired utility boilers to limit their emission of sulfur dioxide (SO_2), nitrogen oxide (NO_x), and particulate matter. To comply with these standards, utilities had to begin constructing plants with a significantly greater amount of environmental emissions equipment than seen in earlier eras. These standards required utilities to begin installing flue gas desulfurization equipment, or "scrubbers," to reduce SO_2 and NO_x particulates.[10]

Recent federal legislation, incorporated in the amendments to the Clean Air Act of 1973 (CAAA), went further in reducing the amount of SO_2, NO_x, and particulates (collectively called acid rain emissions) released into the atmosphere. The CAAA limited all electric utility acid rain emissions by the year 2000 to 1990 levels. If a utility emits less than the 1990 level, it receives a credit for the amount of emissions abated. These credits can be banked for future use or sold on the open market to other utilities that may have exceeded the required 1990 emission levels. The result of this legislation, taken in concert with earlier regulatory standards, has been a significant amount of utility expenditures dedicated to emissions abatement.

Federal economic regulation also had a significant impact on the electric power industry. In 1978, Congress passed the National Energy Act of 1978, which was composed of five different statutes: (1) the Public Utilities Regulatory Policy Act (PURPA); (2) the National Energy Tax Act; (3) the National Energy Conservation Policy Act; (4) the Power Plant and Industrial Fuels Act (PPIFA); and (5) the Natural Gas Policy Act. The general purpose of the National Energy Act was to ensure sustained economic growth during a period in which the availability and price of future energy resources were becoming increasingly uncertain. The two major themes of the legislation were: (1) to promote conservation and the use of renewable/alternative energy, and (2) to reduce the country's dependence on foreign oil.[11]

While all aspects of the National Energy Act affected the electric power industry, PURPA was probably the most significant because it was designed to encourage more efficient use of energy through industrial cogeneration. Encouraging the development of industrial cogeneration met the policy goals of efficiency and reliability in different ways. Cogeneration results in greater efficiency by using industrial process steam as a heat sink for an on-site industrial electricity-generating system.[12] Cogeneration reduces thermal discharge and increases the combined efficiency of electricity and process steam production as opposed to producing each separately.

Greater reliability, another policy goal of PURPA, could be met by increasing the overall number of generators that could be called upon to meet any given load. Unit (generator) availability was a concern for many energy planners during the mid to late 1970s. During this period, close to 100 nuclear and coal power plant construction projects were canceled, raising questions about how to meet future load projections.[13] In addition, operational nuclear power plants were highly unreliable and suffered from significant unplanned and forced outages during the late 1970s. Cogeneration served the reliability goals of PURPA by expanding the opportunity for a whole new class of generators to meet electricity load. Policymakers reasoned that if a traditional utility generator became unavailable, load could theoretically be met by an equivalently sized industrial cogeneration unit.

PURPA was composed of six titles.* Title II of PURPA addressed future policy directions for encouraging cogeneration as an energy efficiency and reliability measure. Section 201 of PURPA defined a new type of electric generation entity: a "qualifying facility" (QF). The strict definitions included in Section 201 defined QFs as those that are "... owned by a person not primarily engaged in the generation or sale of electric power."**

* Title I: Retail Regulatory Policies for Electric Utilities; Title II: Certain Federal Energy Regulatory Commission and Department of Energy Authorities; Title III: Retail Policies for Natural Gas Utilities; Title IV: Small Hydroelectric Power Projects; Title V: Crude Oil Transportation Systems; Title VI: Miscellaneous Provisions.

** Public Utilities Regulatory Policies Act, Public Law 95-617, Section 2.

The key provisions of PURPA (Section 210) are threefold. In large part, these provisions were established to address the barriers to cogeneration. The first provision requires utilities to interconnect with QFs and to provide standby, emergency, and interruptible power. The second provision exempts cogenerators from traditional ROR regulation. The third provision provides a guaranteed market for cogenerated power. Under this provision, utilities are required to purchase electricity from a QF at the utilities' avoided cost. This represented a dramatic departure from the typical pricing of a utility's electric purchases, which set purchased power rates at the cost of service from the supplying source. Under PURPA, purchased power rates would be based on the purchaser's rather than the supplier's cost.

After the passage of PURPA, the Federal Energy Regulatory Commission (FERC) began the process of defining the rules under which cogeneration would be supported in the electric power industry. Part of FERC's charge was to define the specific efficiency and ownership restrictions for a QF.* In addition, FERC also defined the incremental utility costs upon which utility buyback rates for cogenerated power would be based. This definition became known as avoided costs or the costs avoided by a utility (in terms of capacity and/or energy costs) which were avoided by the utility from a QF purchase. The quantitative determination of these avoided costs was left to the states.

The years following FERC's promulgation of PURPA rules saw a number of significant legal challenges that created an aura of uncertainty for industrial firms that sought to take advantage of the legislation's new provisions. These uncertainties were removed in the early 1980s by two important U.S. Supreme Court decisions: *FERC v. Mississippi* and *American Paper Institute v. American Electric Power.***

In 1982, the Mississippi Public Service Commission (PSC) brought an action before the U.S. Supreme Court regarding the constitutionality of PURPA. The Mississippi PSC specifically argued that PURPA mandates forcing utilities to purchase cogenerated electricity within state jurisdiction violated the Tenth Amendment and were thereby unconstitutional. The Supreme Court's ruling disagreed with the Mississippi PSC's argument and held that PURPA did not trample on states' rights and was within Congress' power under the Commerce Clause.

In 1983, the Supreme Court went further in supporting PURPA by reversing a lower court ruling that FERC's rules adopting avoided cost pricing for purchased cogenerated electricity were arbitrary and capricious. The Court ruled that FERC had adequately explained why avoided costs (as implemented in rule) were just and reasonable to retail electricity customers and utilities and in the public interest. This decision, in conjunction

* FERC promulgated rules that defined a QF cogenerator as one that must produce 5% of its total energy output as thermal energy. In addition, utility ownership in a cogeneration project must be limited to less than 50%.

** *FERC v. Mississippi*, 456 US 742 (1982) and *American Paper Institute v. American Electric Power*, 461 US 402, 103 S.Ct. 1921 (1983).

with the earlier *Mississippi* decision, removed most legal uncertainties of cogeneration development in the United States.

During and after the legal travails following FERC's rule promulgation, several states began the process of defining methods for avoided cost-based buyback rates. The methods used to determine these rates varied by state. Some of these methods included:[14]

1. Standard Offers: PURPA requires that state regulatory commissions order utilities to set standard offers to cogenerators. These standard offers set a posted going price for all purchases of electricity. Standard offers were designed to reduce administrative and negotiating costs for cogenerators. Regulators are allowed to set these standard offers for capacity purchases of 100 kW or greater. According to a 1990 survey, about 25% of all states use the legal minimum, 25% have no minimum capacity levels for standard offers, and the remaining states use capacity limits ranging between 200 kW to 1 MW.[14]
2. Levelized Rates: Many state regulatory commissions require that utilities take into account the multiyear aspect of long-term buyback contracts with cogenerators by establishing levelized buyback rates. Levelized rates set the long-run avoided costs at a constant level over several years. This method for setting buyback rates has the effect of increasing cash flow to cogenerators in earlier years and reducing cash flows in later years of the contract period. Such a method assists cogenerators that may face capital and risk constraints in their early start-up years.
3. Avoided Cost Methodologies: The methods employed by regulatory commissions in determining avoided costs can have significant impacts on buyback rates for cogenerated power. Many states vary between requiring utilities to use either short-run or long-run avoided (marginal) costs to derive buyback rates for cogenerated power. Short-run marginal costs tend to lower buyback rates, since they usually only include the short-run operating, maintenance, and fuel costs on marginal units. Long-run marginal costs are typically higher, since they could potentially include the capital costs associated with bringing a capacity addition on-line.[*1] Avoided costs methodologies based upon long-run marginal costs are typically more favorable for cogenerators.
4. Capacity Payments: Cogenerators have the potential to defer new utility generating capacity. For instance, a cogenerator signing a 30-year contract with a utility may defer a planned utility plant for 2 years based upon existing load forecasts. If such a situation occurred, some states would allow the cogenerator a capacity payment equal

* This generalization about short- and long-run marginal costs would be true for a utility experiencing moderate load growth. The higher the load growth, the greater the need for capacity additions, thereby driving up both short- and long-run marginal costs.

to the net present value of the cash savings generated by the cogeneration sales.

5. Competitive Bidding: A more recent innovation in determining avoided costs is to open future capacity needs to bids. Utility and nonutility generators provide competing bids in an auction to meet future supply needs.

These cost methodologies had a direct impact on the level of the buyback rate for cogenerated power and, as a result, the level of cogeneration that was brought on-line in any given state. The more generous the avoided cost methodology, the greater the incentive for cogenerated power. These generous methodologies led to a dual incentive for industrial firms considering cogeneration: (1) an energy efficiency incentive and (2) a profit incentive.*[2] As a result, a significant amount of cogenerated power came on-line during the years following the passage of PURPA. By 1994, the amount of nonutility generation had more than doubled.[3]

Implementation of PURPA resulted in a host of new opportunities for cogenerators. Energy efficiency, by lowering overall energy costs, is one significant opportunity for cogenerating firms. Profits, however, represent an additional opportunity for cogenerators, particularly in states where buyback rates through administratively determined standard offer contracts existed. When the profit opportunity exceeds the energy efficiency opportunity, firms have incentives to bring cogenerating units on-line that do not comply with the original spirit of PURPA. These cogenerators, often referred to as PURPA machines, are located in firms that have weak steam needs in their primary production process and are more interested in producing electricity for a profit rather than increasing overall plant efficiencies. Since PURPA requires firms to meet some minimal efficiency standard, profit opportunities give cogenerators incentives to act inefficiently by dumping waste heat (steam) to meet the minimum PURPA efficiency standards.[15]

Utilities also began to gradually realize the potential opportunities of participating as partners in cogeneration projects with industrial firms both within and outside their own respective service territories. By participating as partners, utilities could negotiate a second-best strategy with industrial firms contemplating leaving the utility system. While utilities could lose a significant amount of load through cogeneration, they could offset these losses by earning a return on their portion of the investment, as well as compensation for constructing and/or operating the new cogeneration facility.

Competitive bidding, initially part of the PURPA process in some states, expanded opportunities for cogenerators by providing profit opportunities for meeting utility loads. As these competitive bidding processes were expanded, opportunities for firms whose primary business was in the independent production of electricity also began to arise. These firms represented

* A number of studies have examined the incentive structure for firms contemplating cogeneration including: Joskow (1982, 1984); Joskow and Jones (1983); Fox-Penner (1990a, 1990b); Rose and McDonald (1991); and Dismukes and Kleit (1998).

a new class of electric power generators known as independent power producers (IPPs).

The expansion of electric generating opportunities, first initiated through PURPA-encouraged cogeneration, has changed the definition of nonutility generators, or NUGs. In the past, NUGs and cogenerators were synonymous, since IPPs (had they existed) would have had no legal market for their power.*[4] However, the expansion of electric generation opportunities has expanded the definition of nonutility generation. The opportunities created by PURPA have led to the competitive wholesale markets that dominate the electric power industry today. Without PURPA, the development of independent or merchant power facilities would have been delayed by at least a decade.

1.2.3.2 State public policy initiatives

State policy initiatives influencing the electric power industry have come primarily from state regulatory commissions and have been concentrated in two areas. The first initiative concerns distributing the enormous financial burdens associated with constructing a large number of nuclear power plants. The second initiative concerns increased regulatory commission activism in the utility resource planning process.

Throughout the history of electric utility regulation, a compact has been assumed to exist between the electric utility and its regulators. This regulatory compact is not a specific written agreement or contract, but rather a generally recognized set of rules, laws, legal practices, and traditions that have defined the legitimate expectations of regulators and utilities regarding the regulatory treatment of large capital investments. The compact places the utility under the obligation to provide reliable service without overcapitalizing. In return, the regulator allows the utility to recover its expenses and the opportunity to earn a reasonable return on its investments.

These traditions, however, could not distribute the unexpected financial burdens of constructing large nuclear power plants. Most completed nuclear power plants experienced much higher than anticipated capital costs. In addition, many utilities needed to recover the cost of projects that were eventually canceled. Regulatory commissions had to review the prudence of investments made in both completed and canceled nuclear power projects. In many instances, state regulatory commissions ruled that a significant portion of a utility's investments were imprudent, and the utility was disallowed from receiving electricity rate recovery.

The magnitude of these disallowances is staggering. An unpublished study by Perl (1986)[16] estimated that total disallowances for both canceled and completed nuclear power plants would total $35 billion.[16] This would be equivalent to almost 54% of the industry's total equity in those disallowed

* Franchise agreements would have precluded IPPs from serving a retail customer independently. In addition, prior to PURPA, few, if any, competitive bidding processes for new utility capacity needs existed.

plants and 16% of the entire equity of all privately owned electric utilities in the United States. These disallowances ranged from between 2% of total costs ($7 million for Harris) to 34% of total costs ($2.1 billion for Nine Mile Point 2). Overall, the average disallowance for plants in this figure is 12% of total cost, or an average investment disallowance of a half billion dollars.

The experience with prudence reviews convinced a number of state regulatory commissions that there was a need for greater *ex ante* regulatory involvement in the planning process.[17] During these investigations, the consensus was that regulators should have played a role in the utility planning process to ensure that the utility selected the least-cost portfolio of resources to meet future demand. A regulatory philosophy about how utility planning should be conducted was developed from this view.

The main philosophical question that was debated in the least-cost planning process concerned the definition of utility planning "resources." In the past, these resources were typically confined to bulk power investments because, under the principle of a natural monopoly, it was more efficient for utilities to produce more power rather than less. Least-cost planning (LCP) expanded this definition by first including demand-side resources such as energy conservation and load management programs. This definition was expanded further when nonutility resources, through competitive bidding, were added to the portfolio of potential utility resources.[18]

The second philosophical question to arise in this process was in defining "costs" and who should bear the responsibility of those costs. In particular, LCP expanded the view of costs from the traditional utility (private) costs to societal costs that include the environmental costs associated with utility generation. These costs are often referred to as environmental externalities, since they are debatably not completely paid for by the party that creates them — thus, the costs are external to the cost-causing party.

Several regulatory commissions, while recognizing the societal costs of electric generation, did not go so far as to require utilities to incorporate these costs into their planning process. Other "external" costs that were considered by regulatory commissions included such factors as risk and uncertainty, fuel diversity, and economic development.

The idea of greater regulatory participation in utility planning was stimulated by a desire to avoid the mishaps of the prudence reviews of the late 1970s and early 1980s.[19] However, the diversity of this new paradigm led to an eventual name change early in the process. The consideration of numerous factors, from stakeholder positions and public interaction to environmental and social implications of utility planning, led the planning paradigm to be called integrated resource planning, or IRP.

IRP is based upon two general principles. The first recognizes that, because of a utility's position as a franchised monopoly provider of electricity, the resource planning process should be open and encourage interaction with a wide range of stakeholders. These stakeholders include regulators, ratepayers, utilities, environmental advocates, IPPs, state planning agencies, and any other entity whose interests are affected by the regulated utility. The second principle is the notion that the utility planning process should consider a wide range of resources and the impacts that these resources have on the environment. These resources include those on both the demand and supply side of the customer's meter.*[5]

One of the main reasons IRP was facilitated as a method to encourage greater deployment of energy efficiency was the belief that utilities had disincentives to promote energy conservation measures as a resource equal to generation. Utilities, for instance, make money by selling electricity. Lower sales through conservation lead to lower revenues and lower profits. IRP, coupled with policies of revenue neutrality, were thought to help remove this disincentive, as well as provide some regulatory oversight to balance the public interest considerations imbedded in utility planning.

Another reason for promoting energy efficiency was the belief that there were a number of market barriers to their utilization by end users. Utilities, with the help of regulatory programs, planning participation, and incentives, were thought to be in the best position to help remove or overcome these barriers. Some of these barriers include the following:

1. High Information or Search Costs: The costs of identifying energy-efficient products or services and their operating/cost characteristics. They include uncertainty about the future benefits and payoffs of these measures.
2. Transactions Costs: Indirect costs of facilitating energy-efficiency opportunities include search time, installation, and even regulation.
3. Access to Financing: Many conservation measures can have significant up-front capital costs. In addition, many lending agencies fail to take into account the income-enhancing opportunities associated with households spending less money on energy, and therefore being able to service loans on energy-efficiency equipment.
4. External Costs and Benefits: A number of external costs are associated with energy production and consumption that are not reflected in price or the utility cost of service. These would include failures to establish time of use and seasonal rates for customers, as well as environmental "adders" to utility generation decisions such as accounting for external benefits of energy conservation that tend to reduce societal costs, i.e., additional employment opportunities, reduced air emissions, and reduced dependence on foreign sources of energy.

Because of the perceived benefits of IRP, the idea took off like policy wildfire through the mid to late 1980s. This effort was sustained by state regulators and the federal government. The research effort promoting

* Demand-side resources include conservation, load management, and cogeneration measures that could be implemented or promoted by a utility. Supply-side resources include building traditional generation facilities, as well as purchased power arrangements with other utilities or nonutility providers of electricity.

LCP/IRP was pursued vigorously by federal energy laboratories, particularly Oak Ridge National Laboratory and Lawrence Berkeley National Laboratory. Funded in part through grants by the U.S. Department of Energy, a Least-Cost Utility Planning Program began with an annual budget of $1 million.[20] The culmination of federal acceptance of IRP was personified in Section 111 of the Energy Policy Act of 1992, which required state regulatory commissions to consider using the paradigm or developing rules that were consistent with IRP principles.

IRP at the state level was pursued to varying degrees. Some states fully embraced the concept, some states rejected the idea outright, and other states opted to choose what they believed to be a hybrid version of the idea. A commonly cited survey of IRP adoption published in 1992 found that 14 states had fully embraced all of the commonly accepted principles of IRP, while another six states were close to full acceptance, missing only one or two general principles of the regulatory paradigm.[21]

The promotion of IRP was strong throughout the 1980s and into the early 1990s. However, starting in 1994, the retail restructuring process and debate began in a number of states, with California taking the lead. In addition, FERC began the process of adopting rules that would eventually lead to Order 888, which required all utilities to provide open and nondiscriminatory access to their transmission systems. The process of wholesale and retail competition, along with the rapid industry change during the period, brought the IRP process to a screeching halt and forced many regulatory commissions to reconsider their involvement in the utility planning process. On a forward-going basis, policy started to allow markets, and not the regulatory process, to determine which resources would be developed to meet customer electricity needs.

References

1. Joskow, P.L., Inflation and environmental concern: Structural change in the process of public utility price regulation, *Journal of Law and Economics*, 1973: 17, 298.
2. Averch, H. and Johnson, L.L., Behavior of the firm under regulatory constraint, *American Economic Review* LII, December 1962: 1052.
3. Energy Information Administration, The Changing Structure of the Electric Power Industry 1970–1991, Washington, D.C., U.S. Department of Energy, 1993: 33–34.
4. Energy Information Administration, The Changing Structure of the Electric Power Industry 1970–1991, Washington, D.C., U.S. Department of Energy, 1993: 36.
5. Energy Information Administration, The Changing Structure of the Electric Power Industry 1970–1991, Washington, D.C., U.S. Department of Energy, 1993: 37.
6. Joskow, P.L. and Rose, N.L., The effects of technological change, experience, and environmental regulation on the construction of coal-burning generating units, *Rand Journal of Economics*, 16, Spring 1985: 6.

7. Energy Information Administration, The Changing Structure of the Electric Power Industry 1970–1991, Washington, D.C., U.S. Department of Energy, 1993: 37.
8. Joskow, P.L. and Rose, N.L., The effects of technological change, experience, and environmental regulation on the construction of coal-burning generating units, *Rand Journal of Economics*, 16, Spring 1985: 22.
9. Utility Data Institute, Electric Utility Power Plant Construction Costs, Washington, D.C., Utility Data Institute, 1994: ii.
10. Utility Data Institute, Electric Utility Power Plant Construction Costs, Washington, D.C., Utility Data Institute, 1994: 35.
11. Energy Information Administration, The Changing Structure of the Electric Power Industry 1970–1991, Washington, D.C., U.S. Department of Energy, 1993: 21.
12. Joskow, P.L. and Jones, D.R., The simple economics of industrial cogeneration, *Energy Journal*, 4, 1983: 3.
13. U.S. Department of Energy, Emerging Policy Issues in PURPA Implementation, Washington, D.C., U.S. Government Printing Office, 1986: 1–3.
14. State PURPA implementation methods are taken from a 49-state survey conducted by Peter Fox-Penner, Regulating independent power producers: Lessons of the PURPA approach, *Resources and Energy*, 12, 1990: 117–141.
15. Barclay, P., Gegax, D., and Tschirhart, J., Industrial cogeneration and regulatory policy, *Journal of Regulatory Economics*, 1, 1989: 226.
16. Perl (1986) referenced in Kahn, A.A., *The Economics of Regulation: Principles and Institutions*, 2 Vols., Cambridge, MIT Press, 1988: xxvi.
17. Goldman, C., Hirst, H., and Krause, F., *Least-Cost Planning in the Utility Sector: Progress and Challenges*, Oak Ridge, TN, Oak Ridge National Laboratory, 1989: 1.
18. Hirst, E., *A Good Integrated Resource Plan: Guidelines for Electric Utilities and Regulators*, Oak Ridge, TN, Oak Ridge National Laboratories, 1992: 12–13.
19. Hirst, E., *Regulatory Responsibility for Utility Integrated Resource Planning*, Oak Ridge, TN, Oak Ridge National Laboratory, 1988: 5.
20. Goldman, C., Hirst, E., and Krause, F., *Least-Cost Planning in the Utility Sector: Progress and Challenges*, Oak Ridge, TN, Oak Ridge National Laboratory, 1989: v.
21. Hirst, E., *A Good Integrated Resource Plan: Guidelines for Electric Utilities and Regulators*, Oak Ridge, TN, Oak Ridge National Laboratory, 1992: 2.

chapter two

Restructuring and the transition to more competitive power markets

2.1 The fundamentals and terminology of power industry change

The current changes now under way in the electric power industry are often referred to in different manners. Three terms that are most commonly associated with these changes are wheeling, deregulation, and restructuring. While wheeling and deregulation are important considerations, they do not completely or accurately reflect all of the changes ongoing in the industry. A closer examination of both of these terms provides greater insights into their descriptive limitations.

Wheeling is a term primarily used by power-industry professionals that describes the third-party transportation of power on the behalf of another utility. Because electric power systems throughout the U.S. are integrated, transporting power cannot be done without the approval of neighboring utilities. Philosophically, a wheeling transaction was not seen as a sanctioned responsibility of utilities; instead, it was something that might be accommodated as a discretionary matter. Locally franchised utilities received regulatory approval at the state and local levels to provide service within a designated territory. Providing power outside that service area represented a deviation from the status quo and was often viewed in a less than favorable light.[1]

A number of events changed the nature of interconnected relationships between utilities. The first fundamental shock to these relationships came in 1969 when the great Northeast blackout forced utilities to reexamine their relationships with one another in order to ensure power supply reliability. As a result of the blackout, the industry took preemptive measures* to form

* The move was preemptive in the sense that the industry formed this voluntary organization before policymakers at the federal and state levels had the opportunity to dictate an alternative reliability arrangement.

the National Electric Reliability Council, which exists today as the North American Electric Reliability Council (NERC).* This voluntary organization typically comprised utilities and was created to pool regional power supplies and increase power coordination for reliability purposes.

The energy crisis provided an additional shock to the relationships between utilities, as well as utility and nonutility sources of electric power. The Public Utilities Regulatory Policies Act of 1978 (PURPA) required utilities to interconnect with qualifying facilities and purchase their power at avoided cost. If the utility were unable to purchase this power, it would be required to wheel the power to someone who would. In addition, due to reliability concerns of the late 1970s, PURPA allowed the Federal Energy Regulatory Commission (FERC) to compel utilities to wheel power on behalf of another utility for reliability purposes. In order for FERC to require wheeling, there needed to be proof that the transaction was for reliability purposes, did not have negative competitive implications for the transporting utility, and was in the public interest.

The next significant step that allowed FERC to compel all utilities to provide wheeling came with the passage of the Energy Policy Act of 1992 (EPAct). Section 211 of this act allowed FERC to compel utilities to provide wheeling to other utilities on an open and nondiscriminatory basis. While the original intent appears to have been on a utility-by-utility basis, FERC generalized the policy by requiring all jurisdictional utilities to offer transmission to all qualified requestors on terms and conditions that are comparable to those transmission-owning utilities provide to themselves.[1]

Wheeling is essentially the technical cornerstone of the current changes in the electric power industry. In effect, wheeling is part of industry jargon used to describe the provision of transmission service by one party on the behalf of another. For instance, assume that we are still in the days of vertically integrated and regulated monopolies. Also assume that there are three utilities: Utility A, Utility B, and Utility C. Furthermore, assume that these utilities are linked in a linear fashion — that is, Utility A to Utility B to Utility C. Now, if Utility A has excess capacity and wanted to sell electricity to Utility C, that power would have to pass through the systems of Utility B. In this case, Utility B would be wheeling, or transmitting, power to Utility C on behalf of Utility A.

The problem with wheeling is that there can be a fundamental commercial disincentive for Utility B to transmit power on behalf of Utility A.** Consider a situation where both Utility A and Utility B have excess capacity and are competing for Utility C's load. Both may have economic sources of generation, but Utility B could squeeze Utility A's ability to complete those sales through its unwillingness to wheel or transmit power. In such

* NERC was expanded from "National" to "North American" with the addition of parts of Canada and Mexico to the reliability organization.

** It should be noted that in the past, utilities frequently wheeled power on behalf of other utilities for reliability purposes. These types of reliability-oriented transactions should be considered separate from competitive actions between two entities.

a situation, Utility B would be using its vertical market power, that is, ownership of generation and transmission, to favor its own operations over those of Utility A. In order to make these types of competitive transactions work, some system of providing open and nondiscriminatory service would need to be developed.

Providing wheeling, or third-party transmission service on an open and nondiscriminatory basis allows competitive providers of electricity to move their power freely across utility systems throughout the U.S. Wheeling can be broken into two categories: wholesale wheeling and retail wheeling. Wholesale wheeling refers to the transfer of power to customers who are not end users, such as wheeling power on behalf of an independent power producer (IPP) selling power to a municipal utility. Or, as in the example above, it could mean transmitting power from one utility (Utility A) to another (Utility C).

Retail wheeling, on the other hand, refers to the physical (and contractual) transfer of power to customers that are end users. For instance, a business choosing to be served by another power provider would have to pay a wheeling or transportation fee to its host utility to receive power from another provider. In this example, a nondiscriminatory system of both transmission and distribution access is needed.

Deregulation is another term that is often used to describe changes now occurring in the electric power industry. However, deregulation is a misnomer when it comes to describing the changes that have and are currently taking place in the power industry. Regulation takes many forms within the power industry. Utilities are currently subjected to significant economic, environmental, and safety regulation at the state and federal levels. Proposed changes in the industry do not refer to removing all forms of regulation. As noted earlier, the prices and earnings of vertically integrated utilities have been regulated. The current changes in the industry envision relaxing price and earnings regulation on the generation and energy sales portion of the industry alone. As will be explained in greater detail later, price and earnings regulation will still remain on the transmission and distribution portions of the industry.

Despite the relaxation of price and earnings regulation on the generation sector of the industry, not all economic regulation will be removed. Rather, economic regulation will be transformed from earnings and price regulation to a market oversight function. Some traditional rate of return (ROR) regulation will remain with the monopoly transmission and distribution (T&D) functions. New players serving electricity customers will require regulators to set and enforce certification requirements, as well as minimum standards for quality of service. In addition, regulators will be required to adjudicate service standard and interconnection disputes between competitive providers of electricity and between competitors and regulated distribution companies. Thus, the use of the term "deregulation" is clearly not an appropriate reference to future industry structure.

Perhaps a more appropriate term to describe the changes in the electric power industry is restructuring. Current industry changes envision breaking the industry into different functional components and introducing competitive forces in one sector of the industry, generation and electric services, while leaving the remaining sectors, transmission and distribution, subject to regulation. To understand these changes, it is important to understand the existing, and what will soon become the historic, structure of the industry.

2.2 *Historic and future structure of the electric power industry*

The electric power industry is referred to as "vertically integrated"; that is, generation, transmission, and distribution are all owned and operated by one company. In addition, prices (referred to as rates) are bundled, reflecting the costs of providing service in all three sectors. Figure 2.1 provides a schematic that highlights the current organization of the industry. Today, all three sectors of production are included under one company. The prototypical electric utility generates electricity from fossil-fueled or nuclear-fired generation stations. The power generated from these stations is transmitted over high-voltage power lines, usually at standardized ratings ranging from 115 to 765 kV. Then power is usually "stepped down" to the distribution level (below 115 kV) and delivered to end users.

Power sales, on the other hand, have been broken into two broad classifications: retail and wholesale. Wholesale sales are reflected as being made to customers who are not "end users"; these include other investor-owned utilities, rural electric cooperatives, and municipally run electric utilities. The wholesale market, which was deregulated at the beginning of 1997,* exchanged more than 2.5 billion kWh in 1998. Today, utilities and other wholesale purchasers of electricity can find competitive sources of electricity for their customers. These increased opportunities have resulted in a 2% increase in wholesale power trades, or 40 billion kWh, between 1997 and 1998.

The retail market, on the other hand, is usually comprised of end users. Since most of these end users are within a state, state public utility commis-

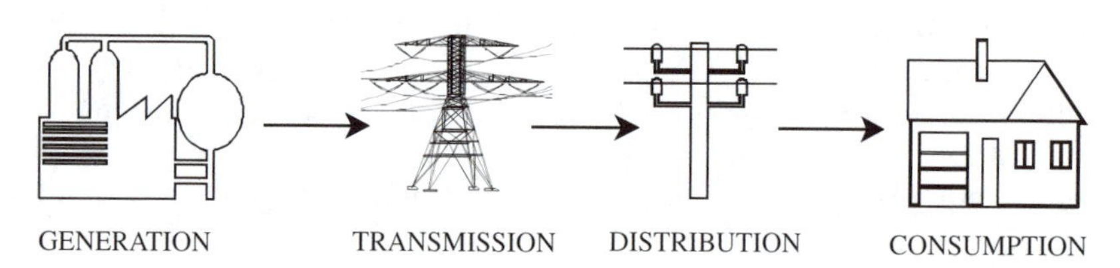

Figure 2.1 Current/historic organization of the electric power industry.

* The Energy Policy Act of 1992 opened the wholesale market to competition. Groundrules for competitive wholesale power exchanges were promulgated by the Federal Energy Regulatory Commission (FERC) in Order 888, issued in April 1996.

sions (PUCs) regulate these rates. The broad classes that comprise the retail sector include residential, commercial, industrial, and other customer classes. The retail market of the industry exchanged more than 3.2 billion kWh in 1998, which amounted to $217 billion in retail sales.

The restructuring of the electric power industry comes from breaking off sections of the industry into competitive and regulated entities. During the past several years, electric generation has become more competitive. As noted in earlier chapters, the fundamental premise of the power generation portion of the business being a natural monopoly does not hold. Hence, competitive forces in the power generation and sales business have resulted in the removal of price regulation. Nevertheless, the lines sectors of the electric power industry (transmission and distribution) are still considered a monopoly.

In the future, competition will govern transactions in electric power markets. Customers will no longer be assigned to utilities and, alternatively, utilities will no longer have guaranteed customers for their electricity. Utilities of the future will compete for end-use customers much like any other good or service. Utilities will be able to contract directly with these customers. Alternatively, middlemen can enter into contracts either linking buyers to sellers (aggregators) or sellers to buyers (marketers) to reduce market informational costs.

Figure 2.2 presents a simple schematic of how the power industry has become restructured at the retail level. On the left side of the figure are a number of power generation facilities and companies. These companies usually sell their power through power (or energy) marketing groups. Their sole purpose is to "market" the output of the facilities. The actual operation, maintenance, and development of these facilities is usually handled through separate affiliated groups within the company.

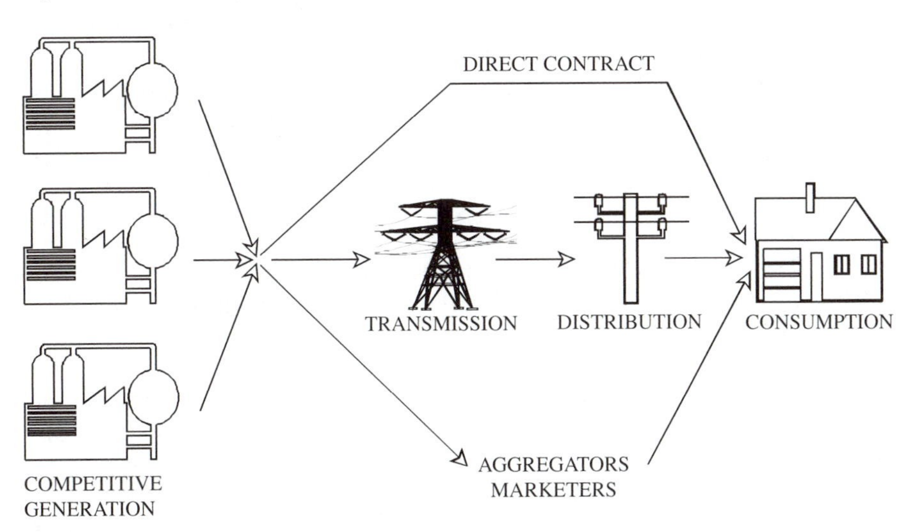

Figure 2.2 Restructured electric power industry.

The figure shows the regulated lines portion of the business as a central medium of physically moving power from one location, referred to as a "source," to a load-serving area, referred to as a "sink." This portion of the industry is regulated. However, the actual sales associated with moving power generation are competitive, and a number of other market participants can attempt to complete these sales with retail customers.

Competitive providers (generators) of electricity can compete for end users directly through their marketing affiliates. There are also market middlemen, such as aggregators and power marketers. Aggregators literally combine end users, or aggregate them, into larger loads to take advantage of their larger purchasing and negotiating power. Power marketers, on the other hand, try to take advantage of arbitrage opportunities between producers of electricity and market prices.

While neither aggregators nor marketers may own power generation, they are important market participants and help keep markets in balance. Both types of participants make their profits through reducing information costs associated with either selling or purchasing electricity. The greater the information costs in a market, the greater the profit opportunities for these types of providers. Both will compete for customers (or selling load) up to the point where the opportunities for gain from information costs have dissipated.

2.3 *The mechanics of restructuring power markets*

While there are myriad issues associated with the move to electric restructuring, there are three basic steps that need to be initiated prior to moving forward. The first step, referred to as unbundling, requires regulators to separate utilities into separate components. The second step requires regulators to establish some form of independence for the transmission system so that open access at the retail level can be facilitated. The third step requires regulators to define the market regime for retail choice. Two polar market regimes define the boundaries for potential retail market structures: the poolco system and the direct access system of choice.

2.3.1 *Unbundling the utility system*

Unbundling utility operations can be a difficult task for regulators, who essentially have to break utilities into generation, transmission, and distribution components. The regulatory goal of this separation is to minimize horizontal and vertical market power of incumbent utilities. If left unchecked, this market power could skew market outcomes in favor of an incumbent utility and significantly reduce the opportunities of societal gains from competition.

Horizontal market power refers to a concentration of assets across one particular business unit or area of production. In the case of electric power, a utility that owned 85% of all generating facilities in a particular market would possess significant horizontal market power. Vertical market power refers to a concentration of vertical assets, such as owning operations in all phases of production from upstream to downstream. Past utility organization (i.e., vertical integration) is an example of a firm that may possess vertical market power. The particular problem with vertical market power arises in a situation where a firm has downstream operations that are monopolistic. In our utility example, this would include a firm that has transmission and distribution affiliates.

To minimize market power of both kinds, regulators will separate utilities by one of two methods: functional unbundling or physical unbundling. Functional unbundling requires utilities to establish functionally separate entities, or subsidiaries, to perform these separate activities. Under a functional divestiture approach, unregulated operations such as power generation, sales, and marketing would be completely separated from regulated operations. Regulated affiliates are required to provide the same information, products, and services to competitors that they do to their unregulated sister companies; no preferences between regulated and unregulated affiliates are allowed.

Physical unbundling requires utilities to literally sell off or physically divest themselves of certain assets. In many instances utilities have been required to sell off a certain portion of their assets, such as a large concentration of generation assets. So, if a utility owned 85% of all generating assets in a given market, regulators could require that this utility sell off some or all of those generating assets to reduce market power problems. In most cases, regulators have asked for the voluntary divestiture of a certain percentage of a certain type of asset (such as all fossil fuel generators). The experiences in California and Texas are good examples of state legislation that has required utilities to sell off a significant portion of their formerly regulated generation assets to reduce market power concerns.

2.3.2 *Transmission independence*

In the future, the transmission system will act as a common carrier. It will transport electricity on behalf of (former) utility and nonutility providers of electricity. However, the terms, conditions, and rates for transmission access are not regulated by state regulatory commissions, but rather by FERC. As discussed earlier, transmission, like generation and distribution, will have to be separated. FERC has not required that utilities physically divest themselves of transmission assets. Thus, the decision to stay in the transmission business is still in the hands of incumbent utilities.

A policy issue not yet completely resolved is the ultimate form of oversight or governance for these transmission assets. Day-to-day operations, operations and maintenance, and long-term planning will have to be conducted by some independent entity that has no ties to the competitive marketplace. How this will be arranged has been an evolutionary process at best. The advent of wholesale and early retail competition saw a strong

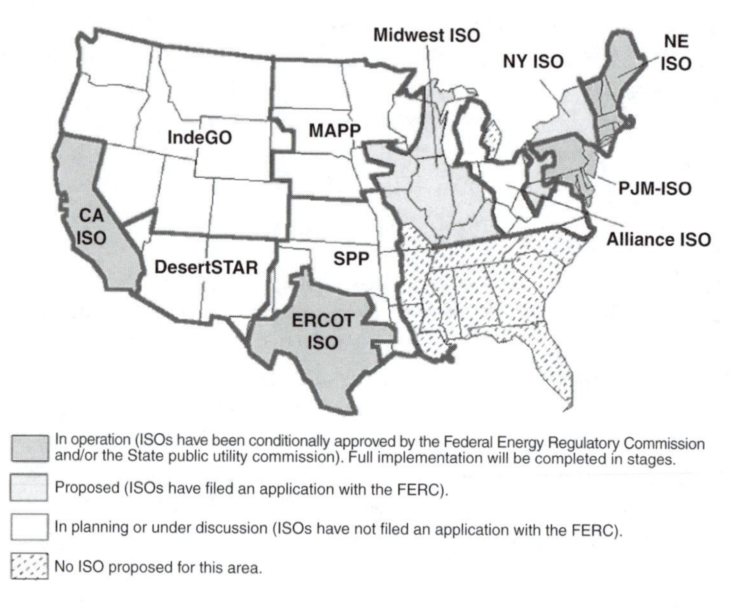

Figure 2.3 Independent system operators in operation, proposed, or under development (March 1998). (From U.S. Department of Energy, Energy Information Administration, Electric Power Annual, 1999.)

preference for the idea of an Independent System Operator (ISO). However, questions about the operating incentives of ISOs have given rise to debates over independent transmission companies, or transcos, and alternative methods for transmission system governance.

ISOs are one of the earliest proposed forms of transmission governance to be facilitated in restructured markets. FERC, in Order 888, gave a strong preference for the ISO concept and its principles. ISOs are essentially non-profit organizations that work like independent air traffic controllers for a given regional transmission system. While ownership of transmission systems stays with utilities, ISOs take over the security and operational control of all power flows and transactions. Neither ISOs nor their employees are allowed to have any financial interest in the transmission system, its operation, or the transactions occurring over the system. An ISO has an independent governing board that includes not only utility representatives, but also representatives from other stakeholder groups, including power marketers, independent power producers, small customers groups, and, in some instances, environmental groups. Recently, the open, objective manner of ISO transmission operation made it a preferred method of transmission organization as seen in Figure 2.3. However, fissures began to develop in this institutional framework and challenged its longer-run viability as an organization structure and paradigm for transmission governance and operation.

ISOs have been plagued by their detractors from the onset of the electric restructuring debate. One initial criticism against the formation of ISOs rested with the enormous costs associated with creating a new bureaucracy to manage regional transmission grids. The experiences and costs associated with the creation of the California ISO and its associated power exchange (PX) provided justification for this criticism. Others argued that the idea of an ISO did not go far enough in removing incentives for cross-dealings and potentially preferential treatment. Other protesters argued that in attempting to develop an open and inclusive governance structure for these ISOs, they had become unwieldy and unworkable. However, one of the most significant criticisms leveled against ISOs rests with concerns about their short- and long-run incentives as nonprofit organizations.

ISO critics have questioned the motivations of nonprofit organizations to efficiently manage and plan for the transmission system. This system will continue to be owned by utilities that have a fiduciary responsibility to their shareholders to maximize the profits that could be earned on these assets. However, a nonprofit organization will be removed from fiduciary responsibility, and may even act at cross-purposes with utility motivations for maximizing shareholder returns. For instance, it is argued that ISOs will have little or no incentives to reduce costs, introduce new technologies, or make management and operating innovations. The inability to earn profits could make ISOs relatively indifferent to long-run planning issues such as increasing transmission capacity or making substation upgrades and additions. The lack of motivating incentives has led many critics, primarily utilities, to call for an alternative transmission governance and organizational arrangement.

One of the more recent proposals for transmission organization rests with an institution/corporation known as a transco, which is short for transmission company. The transco idea attempts to merge the concepts of independence and inclusiveness of an ISO with the profit-maximizing goals of a private enterprise. Recent transco proposals envision a multientity (multiutility or regional) corporation that would operate and manage utility transmission assets. The utility owners of these assets, in turn, would serve as shareholders in this new corporation. Management of a transco would then be accountable to its shareholders. Transcos would be for-profit entities, but would include membership and (nonvoting) input from nontransmission-owning stakeholders such as municipal utilities, rural distribution cooperatives, power marketers, and IPPs. These participants could become voting members of the transco if they decided to buy shares in the newly formed company. These shares would be open to all who wished to purchase them at fair market value.

Critics of the transco proposal question the motivations of the transco. Its operating goals, critics argue, would be in maximizing profits, not in facilitating trade. Critics point to examples where a transco could be faced with the choice to help accommodate trade or block a potential trade to profit from its strategic control of a bottleneck (monopoly) asset. In addition, many are concerned that utility domination of such an organization is highly likely, since most governing arrangements give preference to the decisions and input of transco shareholders, many of whom have significant generation assets.

Thus, while establishing independence for the transmission system is a basic and crucial step in moving forward with electric restructuring, it is a difficult task to accomplish, at least to the satisfaction of all interested parties. The core issue is identifying the motivating goals for the transmission system: should it maximize the number of trades or should it maximize profits? An additional question arises concerning the degree to which these alternative goals are at odds with one another. It is likely that transmission system governance and organization will be an evolutionary process until these two alternative goals are reconciled.

The challenge for federal regulators has been to encourage development of independent organizations and to do so in a manner broad enough in scope to secure independence, as well as potential operating efficiencies across regions. In a recent order, FERC took its boldest stand on the issue by forcing all parties to the table for 45 days of negotiations to organize the U.S. power transmission system into five major systems: West, South, Northeast, Midwest, and Texas. These systems will be organized into large regional transmission organizations (RTOs) that will handle a variety of different transmission operation, pricing, and planning issues. While it is still too early to tell, the promise of having a number of large regional RTOs, with a number of for-profit transcos, seems likely.

2.3.3 Market equilibrium and trading regimes

The goal of electric restructuring is to develop more competitive electric power and generation markets. Competitive markets, as opposed to those that are regulated, are thought to provide better incentives and opportunities to both buyers and sellers. However, in order for restructuring to achieve its most optimistic goals, power markets must become perfectly (or nearly perfectly) competitive. This brings up the question: what makes a competitive market?

Over the years, economists have developed a set of conditions that characterize perfectly competitive markets. These conditions include the following:

1. Homogeneous good
2. A large number of buyers and sellers, each of which are small relative to the market
3. Perfect market entry and exit
4. Sellers are price takers, not price makers, and prices are set at marginal costs

These conditions essentially give both buyers and sellers significant opportunities for mutually beneficial gains from trade and, if these conditions hold, will prevent monopoly market power and monopoly profits from arising. Monopoly market power simply states that a single firm dominates a market and sets prices at a point that is much higher than costs. The presence of economic profits indicates that a firm or firms in a market are earning profits that are above a normal rate of return. The conditions above, if they hold, will prevent either of these results from occurring.

For instance, the condition of a homogeneous good virtually ensures that all buyers have access to perfect substitutes. If the electricity sold by Company A is the same as that sold by Company B, buyers can change service providers if the price offered by one company exceeds the going market rate. This condition ensures that no provider can set a price on a condition other than costs, since the product offered in the market is the same throughout.

The presence of a large number of buyers and sellers also provides markets with some obvious benefits that reduce market concentration and power. If there are few sellers in a market, they have the ability to increase prices, given the lack of competition. In addition to a large number of firms, there exists a corollary condition to perfectly competitive markets that states no one single firm is large relative to the market. This condition prevents one firm from using its size to dominate the market.

The condition of perfect entry and exit in a market allows competitors to freely enter and leave to take advantage of profit opportunities. This disciplines markets. If there are perfect entry opportunities, firms that see economic profits will enter to take advantage of these opportunities. These firms will offer prices marginally lower than those of the incumbent firm or firms to steal away their business. Other firms will enter and offer prices progressively lower. This process continues until all economic profits are bid away through lower prices.

The condition that prices are set to marginal costs is an important one for understanding not only the workings of competitive markets, but also what industry stakeholders hope to see in future restructured electric power markets. Setting prices at costs, in this case marginal costs, ensures that no economic (or monopoly) profits exist. In the case of the electric power industry, marginal costs are determined by the incremental costs of generating electricity.

Assuming that almost all of the above conditions for perfectly competitive markets are satisfied, or nearly satisfied, we can describe how market-clearing prices are determined in electric power markets. Market-clearing prices are generally determined by the intersection of supply and demand; in other words, when we have market equilibrium. This condition is presented in Figure 2.4.

Note that there are two curves represented in Figure 2.4, both of which are plotted against prices and quantity (or electricity). The demand curve is indicated by D and begins in the upper left corner of the graph and slopes downward. The downward slope indicates that as prices fall, quantity demanded increases. This is simply a graphic representation of what economists refer to as the first law of demand.

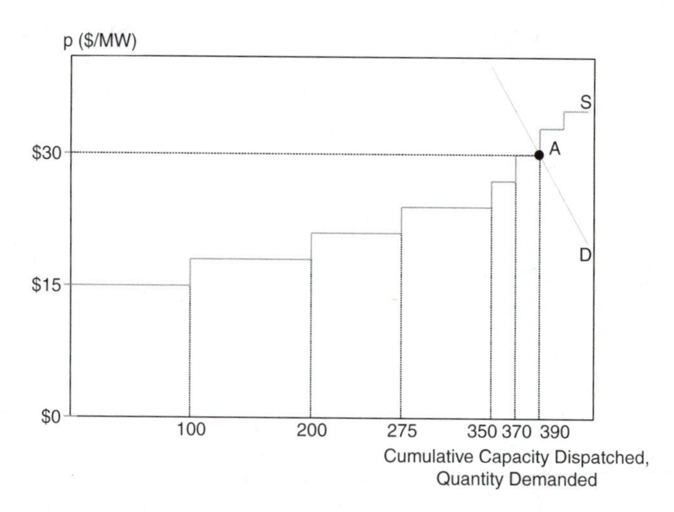

Figure 2.4 Power market supply, demand, and market equilibrium.

Within the same graph we have a representation of a supply curve, indicated by S. This curve starts at the bottom left corner of the graph and slopes upward. The supply curve simply provides a graphic representation of the fact that as price increases quantity supplied, dispatched electrical capacity (or power plants) increases. Prices have to increase given the increased costs of providing an additional unit of electricity.

The intersection of the supply and demand curves is known as the market equilibrium and defines the conditions necessary to determine a market-clearing price. This equilibrium point is indicated by Point *A* in the graph and represents a price/quantity combination at which both buyers and sellers are mutually better off by trading. Market equilibrium prices are given by $30/MW on the vertical axis, and market equilibrium quantity traded is given by point 390 MW on the horizontal axis. The equilibrium point is said to be efficient because society is making the best possible use of its resources.

The example in Figure 2.4 presents a simple illustration of how power markets work. The condition that determines the equilibrium is true regardless of whether we are talking about short-term spot markets, which are usually defined on an hourly basis, or longer-term market transactions. Most discussions of competitive markets, however, focus on using this example for market-clearing conditions in spot markets. The result, however, can still be generalized to longer-term transactions and markets.

With these preliminaries in hand, we can discuss two polar market structures that have been proposed for restructured electric power markets: the poolco structure (essentially a nonprofit "pooling company") and the bilateral contracts, or direct access, market structure. These market models represent the extremes of potential market organization in a restructured environment. The poolco market structure can be thought of as a centralized trading regime that attempts to maximize wholesale market gains. The bilateral contracts market structure, on the other hand, is very decentralized and allows end users full access to suppliers and competitive providers of electricity.

The poolco market structure essentially attempts to convert the retail market into a large wholesale trading regime. Most proponents of the poolco structure would argue that a relatively large state or region is needed to tap the opportunities in such a structure. While variations on organization of the poolco do exist, there are essentially two functions. First, the poolco serves as a market-clearing institution, or power exchange (PX), that brings together regional buyers and sellers of electricity. Second, the poolco serves as a coordinator of power across regional transmission grids and as an institution that manages day-to-day operation and maintenance of the grid and provides long-term planning. In other words, the poolco serves as the ISO.

The poolco is essentially a large wholesale trading/coordination entity. All competitive generators and suppliers would be required to bid, on an hourly basis, the amount of power and price they are willing to offer. The PX ranks these price/quantity offers on a least-cost basis, much like prior arrangements under tight power pools that have existed in the past. The proposed least-cost dispatch is then passed to the ISO for coordination purposes. If conflicts arise, the ISO works with the PX, who in turn works with suppliers to develop alternative dispatch configurations should a congestion problem arise.

Under this structure, regulated distribution companies (discos) would be required to purchase their electricity from a poolco. All power is dispatched by the poolco and all demand is met. The cost/price offered by the last dispatched unit, which is referred to as the marginal unit, sets the market-clearing price. All generators that have been dispatched prior to the marginal unit are paid that unit's offered price. The lower their cost relative to the marginal unit's dispatched cost, the higher that individual generator/supplier's profit.

On the other extreme of the market structure spectrum is the bilateral contract, or direct-access trading regime. Under a direct-access regime, customers individually negotiate with sellers of electricity. This market works very much like other goods and services markets. An analogy can be drawn to long-distance communications, where customers have the choice to leave their existing provider of long-distance service for an alternative competitive provider. An unhappy customer can switch back to the incumbent provider or another competitive carrier.

The two market structure regimes raise the question about which structure is the best. Certainly, this is a policy question; both structures have perceived benefits and costs. For instance, the poolco market structure helps to achieve substantial gains in wholesale markets first and is a relatively quick way to move forward with a limited form of competition. Its disadvantage is that it treats electricity as a plain-vanilla commodity and does not really facilitate the service and quality opportunities that have been envisioned for restructured markets. The bilateral contracts market, on the other hand, helps to customize electric service needs of customers and facilitate service diversification on behalf of suppliers. However, such a system is perceived as being inequitable because smaller customer classes tend to get lower price discounts than do large bulk customers.

The answer to the market structure debate is more than likely somewhere in the middle of the two market structure regimes; close examination reveals that they are not mutually exclusive. For instance, both structures require a functional unbundling of incumbent utility operations. Both structures require (or suggest) the creation of an ISO. Both markets facilitate market-driven, rather than regulatory, pricing. The creation of a PX as a market-clearing institution could be thought of as a short-run mechanism to facilitate choice and to serve as a backup institution for retail providers of last resort. For this reason some states have seen the poolco structure as a first step toward the eventual evolution of a full-scale direct access/bilateral contracts market.

Reference

1. Fox-Penner, P., *Electric Utility Restructuring: A Guide to the Competitive Era*, Vienna, VA, Public Utility Reports, 1998: 158.

chapter three

The first major challenges to the system: the California restructuring experience

3.1 Introduction

The passage of the Energy Policy Act of 1992 (EPAct) began the process of opening electric power markets to competition. Soon after its passage, California became one of the first states to examine, adopt, and implement electric restructuring at the retail level. At the time, California was considered the "poster boy" for opening retail markets to customer choice and competition. However, as was soon to be seen, the devil of restructuring was in the details, and the California approach was filled with them. As a result, the once-touted state for restructuring has now become one that is disparaged for having moved too quickly, in too much detail, on an issue that many see as being too complex.

The problems with California power markets were amplified during the summer of 2000 and through the better part of 2001, when prices in the state's wholesale power markets surged to unprecedented levels. These price increases were immediately felt by some of the state's ratepayers, since price cap protections, initially established to protect customers during the transition period, were removed. As a result, customers in the greater San Diego area saw their bills more than double their already nationally high levels. Customers in other regions of the state, on the other hand, will probably only be temporarily sheltered from these significant rate increases, since the remaining price-capped utilities in the state are already calling for recovery of the additional costs of making purchases on the wholesale market.

The current events in California have government regulators and policymakers at all levels scrambling to assure their constituents. Committee meetings are being held, investigations are being conducted, and even the courts may see some of the action from these events. The two most pervasive claims that have been made to date about the current problems of the market have been:

1. These markets are being controlled by competitors that are strategically manipulating bids into the state's power exchanges.
2. The industry, and customers, are not ready for competition at this point, and a retrenchment to some sort of regulatory control may be in order.

Neither of these claims squares with the experiences or trends in broader electric power markets where competition has garnered some reasonable degree of success. Nor does the policy proscription of moving back to regulated markets make any sense — regardless of the reasons for the current crisis. The genie of competition is out of the bottle, throughout the United States and internationally; putting it back is even beyond the ability of a state as large and progressive as California.

Perhaps the true problems with the wholesale market in California rest with:

1. The complicated interaction of supply and demand in an evolving market
2. The existence of a complicated and centralized market structure that has limited the opportunities for buyers and sellers to interact openly and freely

A number of issues related to these themes, believed to be supported by a more objective analysis of the situation such as the one recently presented by the staff of the Federal Energy Regulatory Commission (FERC),[1] will be explored.

3.2 *Background on the creation of the competitive California market*

Prior to 1997, California's power market was structured much like other retail power markets around the country. Generation, transmission, and distribution were owned, operated, and priced on a bundled basis. The state was dominated by three large investor-owned utilities (IOUs) that included: Pacific Gas & Electric (PGE); Southern California Edison (SCE); and San Diego Gas & Electric (SDG&E).

Before the advent of electric restructuring, California had a reputation for pursuing the latest innovations in utility regulation and planning. In part, the motivation for this progressive nature came as a result of the general nationwide trend toward more detailed oversight and planning by regulatory commissions through the integrated resource planning process (IRP). This IRP planning philosophy was adopted wholeheartedly by California in the belief that utility regulation and planning should consider a wide range of utility planning implications, including the environment, the economy, and social goals. Thus, policies like demand-side management, renewable set-asides, standard-offer contracts for qualifying facility (QF) power, revenue neutrality and decoupling, and incentive returns on energy conservation programs became the vogue in California regulatory design.

In addition to the state's progressive nature in regulatory policy, there was a veritable cadre of stakeholders that intervened in the regulatory process in California. All wanted their voices to be heard and their special programs to be continued. These intervention groups ranged from consumer advocacy groups to environmental groups to utility shareholder groups, low-income groups, and utility worker unions, to name a few. The California regulatory process, to its credit, was open and encouraged this activism. However, the development of regulatory policy over the years rested on the belief that all of these different interests could be accommodated to one degree or another.

This expansive precedent for regulatory activism is the backdrop for electric restructuring in California. Starting in 1995, the California Public Utilities Commission (CPUC) adopted a general approach for opening the state's retail markets for competition. The debate in these proceedings heralded the two opposing market structure paradigms: a bilateral market approach vs. a centralized pool, or "poolco," approach. Eventually, the poolco model became the marginally preferred approach. Soon following the 1995 ruling, the commission moved forward with proceedings to discuss how the actual poolco approach would be implemented.

However, within a year, the California Assembly decided that it would like to get involved in the electric restructuring debate. As a result, AB 1890 was passed and signed into law by Governor Pete Wilson on September 23, 1996.[2] The bill, in true California fashion, had a number of perks for stakeholder groups engaged in this debate. For the IOUs, there was a guarantee for stranded cost recovery. For the ratepayer advocates, a 20% rate reduction and rate freeze were offered as long as the utilities continued to collect their surcharges, known as competitive transition charges (CTCs). For the environmentalists and other public interest groups, long-support public programs were guaranteed to be maintained in the near future through other bill surcharges and collections. The statute codified the poolco approach into law and put the wheels in motion to create two large, nonprofit power institutions: the California Independent System Operator (Cal-ISO or ISO) and the California Power Exchange (CalPX or PX).

The Cal-ISO was created to independently operate 75% of the state's extensive transmission system. The California transmission network consists of 21,000 circuit miles of power lines that deliver about 165 billion kilowatt-hours (kWh) of electricity every year.[1] Power plants connected to the Cal-ISO have a total capacity of approximately 45,000 megawatts (MW).

The goals for the Cal-ISO are to run the state's transmission system like an independent air traffic controller. The ISO's primary responsibility is the physical operation and planning for the transmission system. The ISO's specific responsibilities in the new market include:

1. Maintaining open and equal access to the transmission system for all potential participants
2. Procuring ancillary services to maintain reliable operations
3. Managing day-ahead and hour-ahead schedules
4. Performing real-time balancing of load and generation
5. Settling real-time imbalances

The other institution dominating California's wholesale market is the CalPX. This institution is responsible for organizing and clearing the commercial transactions of the system. The CalPX conducts daily auctions for competitive supply of the state's markets. CalPX accepts demand and generation bids (price, quantity) from its participants, and determines the market-clearing price (MCP) at which energy is bought and sold.[2]

The CalPX also serves as the market analog to the ISO and assists the ISO to alleviate operational constraints through market mechanisms such as establishing:

1. Ancillary service bids to maintain system reliability
2. Adjustment bids, which are incremental and decremental bids to alleviate congestion problems on the transmission grid
3. Supplemental energy bids, which are used by the ISO to match loads and resources on a real-time basis[2]

While both institutions control wholesale markets, their lines to retail markets are direct. All power that is sold to retail end users must be coordinated and scheduled through California's power institutions. Retail customers choosing alternative providers must at some point deal with scheduling coordinators that submit balanced supply and demand schedules to the ISO. Other customers taking default power pay for power secured from the ISO at prices that clear within the CalPX.

While the above discussion seems like an esoteric description of issues associated with power markets that most customers are either unaware of or uninterested in, they have important implications for the past and present crisis in California. The current operation of the California market has its precedents in the past structure of regulation within the state.

As noted earlier, California has a historical inclination to adopt relatively innovative and complicated policies associated with utility regulation and planning. The discussion above shows that the development of restructured power markets in California has not deviated from this trend. It is the temptation to micromanage markets in California that has created a large part of the problems the state is experiencing today.

3.3 The capacity availability dilemma

Electricity is a unique commodity in the sense that it cannot be stored and must be produced simultaneously with demand. In the past, and even into the present, generation planning has consisted of developing and having a portfolio of generation facilities available to meet the various types of electricity loads that occurred in any given hour, across any given day, in any given season. In the past, reliability tended to be the most important planning consideration, followed closely by cost. Thus, generation planning strategies consisted of constructing and operating enough power plants to meet demand on a cost-effective basis. In many instances, having the ability to meet sudden surges in demand entailed constructing and maintaining large capacity reserve margins that remained idle during large parts of the year. In the past, regulators determined the degree of reliability. Today, that degree of reliability is determined in large part by the market.

Since load varies considerably across hour, day, and year, the power industry has traditionally recognized three different classifications for power facilities: baseload generation, intermediate or cycling generation, and peaking generation. Baseload generators are typically steam generation facilities that are used to service minimum system load and, as such, are run at a continuous rate. While these units are the most efficient to operate, they are costly to start up from a cold shut down, therefore, they are usually run at a near-constant rate. Intermediate load plants are typically older steam units or combustion turbines that are brought online during periods of forced or planned outage of baseload units. Intermediate units can also be thought of as units that bridge the dispatch of baseload and peaking units during periods of unusually high demand. These units can be older and are less efficient than baseload units. Peaking units are typically combustion turbines that have the ability to generate electricity immediately and serve temporary spikes in demand, such as during a heat wave when residential and commercial air conditioning demands begin to surge.

In the past, electric utilities dispatched generating units to meet demand on a lowest- to highest-cost basis. This form of dispatch is commonly referred to as "economic dispatch." The marginal or incremental cost of dispatching units is traditionally the benchmark used to rank order available generators. These marginal costs, in the very short run, are typically associated with changes in fuel costs and other variable operating and maintenance (O&M) costs. Historically, baseload units, almost always large coal, hydro, or nuclear units, had the lowest incremental costs and were dispatched first to meet load. As load increases during the day, or across seasons, less efficient intermediate or cycling units, which generate electricity at slightly higher costs, were brought online. Higher-cost peaking units would be the last types of units brought online under an economic dispatch regime. The cost of the last dispatched unit therefore defines the system marginal costs, often referred to as the system "lambda."

Power plant dispatch in the California model works much like it did under the days of utility regulation, with some differences. For instance, generators "bid" to the CalPX for the right to be dispatched, whereas in the past, this function was completely internalized within a vertically integrated utility. The system lambda today, for the CalPX, can be thought of as the market-clearing price, or MCP. In addition, like the operational practices of the past, the Cal-ISO uses peaking units to meet temporary and sudden surges in demand. These plants are traditionally smaller, have relatively low capital costs, but high operating costs, since they are fossil-fuel-fired and of low efficiencies. These plants are usually dispatched quickly to meet demand at very short notice.

Capacity margins are the metric by which a power market's tightness is measured. Capacity margins are defined as the difference between the total power plant availability measured in MWs, less peak demand, divided by total generating capacity; in other words, the excess capacity in the market as a percent of total capacity. As can be seen from Figure 3.1, these historic margins have been relatively high and, traditionally, an 18 to 22% margin was considered to be part of a prudent planning and reliability strategy.

A number of events during the past 10 years have changed the industry's historically high capacity margins. First, economic growth around most of the country has been strong and has whittled away a lot of the excess capacity that accumulated during the 1980s. Second, regulatory policy prior to restructuring did not provide a strong environment to encourage utilities to build new generation. During the period from 1988 to 1994, the industry, still reeling from the prudence disallowance experience, was encouraged to rely on demand-side alternatives as opposed to constructing supply-side power plants. After 1994, utilities became concerned about the potential regulatory treatment of power plants as the clarion call to electric competition became louder. Thus, regulatory uncertainty prevented a number of traditional utility resources from coming online during the 1990s.

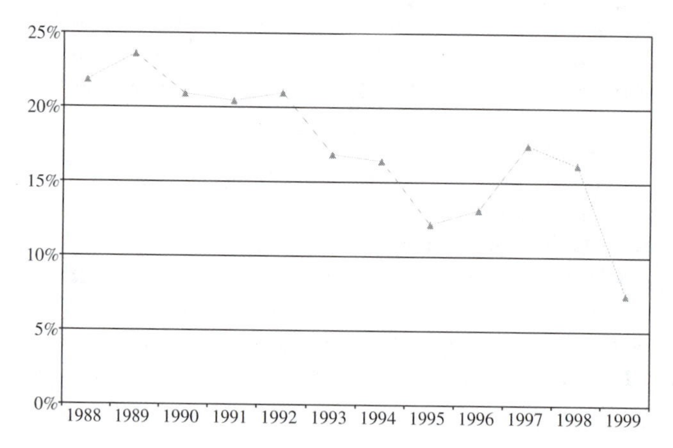

Figure 3.1 Historic U.S. capacity margins.

The combination of these two factors has resulted in precariously low capacity margins as California leaped into retail choice virtually overnight. During the past 17 years, the Western Systems Coordinating Council (WSCC), which encompasses the entire western portion of the United States, has been experiencing substantial growth in peak demand. The annual average growth in peak demand for California during this period (1982–1998) was approximately 3.2%, compared to an annual average increase in generating capacity of less than 1%.[1]

Competitive markets responded to the short supply of generation resources by using price to ration the market. The unfortunate bystander in this experience has been consumers. For customers in San Diego, where price caps were removed due to the early retirement of SDG&E's stranded cost, these costs were directly passed through on monthly bills. The customers of SCE and PGE have, in the short run, been protected, but both utilities are clamoring that someone (i.e., ratepayers) make up the differential between capped rates and high-priced PX purchases. If the regulatory policies regarding stranded costs are any indicator, then it is just a matter of time before California's ratepayers get stuck with the bill.

Most observers of competitive markets would suggest that high prices send signals to entrepreneurs to make investments. In fact, this is occurring to a large extent throughout the United States today in electrical power markets. Figure 3.2 shows the number of announced competitive merchant power plants in the United States. The shaded states are those that have, or are moving toward, electrical restructuring. The combination of open markets and capacity shortfalls has sent the appropriate signals.

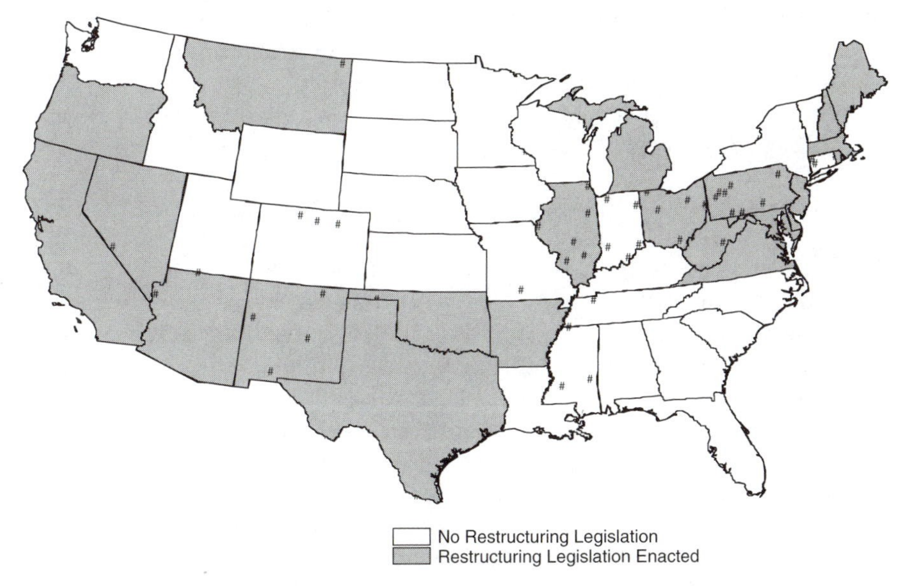

Figure 3.2 Merchant power plants coming online prior to 2000.

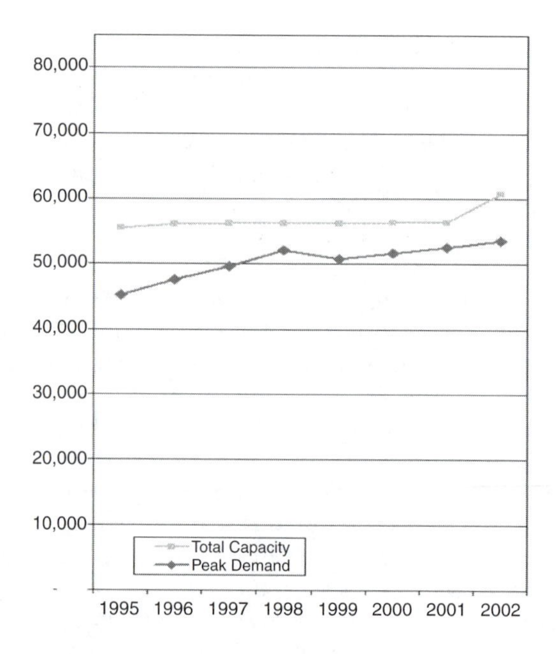

Figure 3.3 Total capacity and peak demand in California.

One of the exceptions to this rule, however, rests with California, where there has been little merchant activity. The reason? Environmental regulations that are both stringent and time consuming for merchant developers. While no one supports plants that unnecessarily pollute the environment, the consequences of these regulations will prevent the state's capacity margins from growing anytime in the near future. Figure 3.3 shows the forecasts for demand and supply growth over the next several years. Tight margins will continue to produce high prices until the state's environmental priorities are reconciled with its power resource needs.

3.3.1 Thin market generation ownership

The state's generation market limitation does not rest with size alone. Clearly, the state could do more to encourage a larger number of new competitors in the market. Today, IOUs control only about 20% of the total generating assets in the California market, while nonutility generators (NUGs) hold around 48%. At first blush, this would appear to represent a relatively competitive mix of generating assets. However, while California law strongly encouraged the state's IOUs to divest themselves of a good portion of their nonnuclear generation assets, no new growth of generation resources occurred. Thus, the restructuring of industry in California consisted of a lot of shuffling of existing resources without the significant growth of new players.

Table 3.1 Comparison of PX Supply Relative to Capped Market Prices

Price Cap Level (MW)	Cal IOUs	New Generation Owners	Power Marketers
$250	18,018	4445	3506
$500	19,270	5277	3615
$750	18,514	4441	3483
On-Peak Total	18,544	4647	3523

Source: Federal Energy Regulatory Commission on Western Markets and the Causes of the Summer 2000 Price Abnormalities Staff Report, Washington, D.C., November 1, 2000, p. 3–25.

This problem becomes more confounded when one considers that a nontrivial portion of the state's generation mix is dedicated to what are known as regulatory must take (RMT) units. These RMT units are those that have been defined by the CPUC as subject to specifically identified cost-recovery contracts. These resources are scheduled through the CalPX and include such facilities as nuclear power plants, QF facilities, and long-term purchased power agreements signed by utilities under the days of traditional rate-of-return (ROR) regulation. These resources are usually "zero-bid" into the PX to ensure their (economic) dispatch. Thus, during certain hours — especially nonpeak hours — RMT units could take up a significant share of the dispatched capacity, thereby dissipating any diversity in generation ownership that may be apparent from viewing capacity ownership overall.[7]

The problems associated with generation availability by ownership type are exacerbated when compared to the price-cap mechanisms imposed on the system during the recent crises. For instance, between August and June of 2000, the Cal-ISO imposed a cap on prices starting at $750/MW, which was eventually reduced to $250/MW. During this time period, the proportion of electricity supplied by IOUs was actually greater. Table 3.1 presents a breakdown of CalPX on-peak supply bids under different ISO price caps for selected participant types.

3.4 *Transactional limitations for buyers and sellers*

Some of the other problems associated with California's competitive power markets rest with the restrictions on contracts that can be acquired by large purchases of electricity — particularly the large distribution companies of the IOUs. Restructuring processes in California require all purchasers to secure their power from the CalPX. Until recently, there were very few purchasing options. These consisted of the spot market (day of) or forward market, which consisted primarily of day-ahead market.

Forcing distribution companies into these two types of transactions seriously limits the types of risk management activities that could be facilitated. Hourly markets, for instance, are clearly going to be more volatile than longer-run contracts. These longer-run contracts allow parties to negotiate risks and pay premiums in return for some degree of limited price exposure.

There is no reason why the full brunt of the increases in prices from the PX needs to be directly passed on to customers. For instance, mechanisms could have been developed whereby utilities offer customers the option of more stable prices throughout the year by seasonal averaging (something SDG&E has since added), or alternatively through a wide array of other options that could include long-term bilateral contracts, future and forward contracts, and other derivatives. However, development of these mechanisms was not only unavailable, it was discouraged so as to not risk the undermining of the liquidity of the newly created PX. Until recently, the CPUC placed strict limits on the options available to IOUs to enter into these forward markets. These limitations included restricting any forward contracts to be limited to no greater than 12 months.[1] In addition, the CPUC limited the amount of energy that could be purchased under these forward contracts.[1]

Despite these limitations, the FERC report notes that in the course of its investigation, the IOUs in California did not utilize even the limited amounts of opportunities they had to enter into forward contracts. The report speculates on the potential reasons for this. First, the standard products offered by the PX over the period in question may not have met, or were not perceived to have met, the needs of the IOUs. Second, because the standard-offered forward contracts did not provide a full range of hedging features, they may not have offered the level of insurance against price spikes that the IOUs needed. Third, the prices for the block-forward contracts may have been high relative to the IOUs forecasts.

While all of these seem to be plausible results, two facts remain. First, the IOUs did not attempt to encourage either the PX or the CPUC to develop new opportunities, instruments, and contracts for the forwards market, even when the crises were developing. Second, very few instruments were facilitated anyway. The problem, to a cynical observer, may be that since the probability of passing along excessive prices was high, there was no incentive to reduce costs by reducing risk for the IOUs on behalf of their captive customers. The situation is similar to the days of traditional regulation when utilities were allowed to flow through fuel charges for power generation to end users. The incentive to minimize costs in such instances is decreased when there are no downside risks to utilities.

3.5 *Failure of analysis*

One of the last points to make about current events in California markets has to deal with a failure in their basic analysis. The current conventional wisdom regarding California power markets is that the state's new merchant competitors are taking advantage of some unfortunate structural problems in the system. These critics claim that the costs of generating a MWh cannot equal $1000, compared to a cost that, at the worst, could equal $200. In fact,

Table 3.2 Correlations of Western Market Prices: On-peak Prices from Megawatt Daily and California Power Exchange (May 1–August 21, 2000)

	COB	Mid-Columbia	Palo Verde	4 Corners	NP 15	SP 15	Cal PX NP 15	Cal PX SP 15
COB/NOB	1.000							
Mid-Columbia	0.997	1.000						
Palo Verde	0.971	0.963	1.000					
4 Corners	0.961	0.953	0.995	1.000				
NP 15	0.992	0.987	0.974	0.966	1.000			
SP 15	0.969	0.960	0.992	0.983	0.977	1.000		
CalPX NP 15	0.912	0.908	0.865	0.858	0.919	0.876	1.000	
CalPX SP 15	0.915	0.906	0.932	0.932	0.922	0.937	0.930	1.000

Source: Federal Energy Regulatory Commission on Western Markets and the Causes of the Summer 2000 Price Abnormalities Staff Report, Washington, D.C., November 1, 2000, p. 3–16.

even the FERC staff analysis makes some comparison of accounting costs to market to make some generalized conclusions about market power.*[1]

What these pundits fail to realize is that all markets in the United States are tight, and this has been especially true in the broader western markets. The open access provisions created by EPAct, and promulgated by FERC in Order 888, have created regional, and even national, opportunities for selling wholesale power. Thus, power sold in California will have to compete with power from neighboring states and regions. If resources move to where their returns are most dear, then it should come as no surprise that high surges in California PX prices move in the same direction as those nearby areas. Thus, the problem with California's market may have less to do with what is going on within the boundaries of the state as it does with what is occurring in the entire region.

An indicator of the relationships between California markets and other regional trading hubs can be reflected in the correlation statistics between the various markets. Table 3.2 presents the correlations between the various western power hubs, including: California–Oregon Border (COB); Palo Verde; 4 Corners; Mid-Columbia (Mid-C); North Pass 15 (NP 15); and South Pass 15 (SP 15). The table shows that correlations among these markets are quite strong and statistically significant. Correlations among all of the western on-peak prices show correlations of 0.858 or above.

* The FERC staff report concluded that market power may exist because of the huge mark-ups above marginal costs that many generators garnered in the market. However, the report was very clear in stating that it was unable to attribute any strategic behavior on the part of any individual — or any class — or market participants during the summer price run up. The result is interesting because it is contradicted by the other analysis included in the report that shows strong correlation in regional prices — indicating a competitive regional market.

3.6 Conclusions

The situation in California is an unfortunate turn of events for advocates of electric restructuring. The crises in western power markets have shaken the confidence of many supportive, but questioning, policymakers across the country. It has also poured fuel on the fires of those interests that are adamantly against moving forward with more competitive retail markets. In states from New Mexico to Louisiana, policymakers have reconsidered — or stalled — restructuring initiatives that have been ongoing for several years.

However, there is a silver lining from these events: the lesson that can be learned from California. Clearly, one of the best ways to open competitive retail markets is to encourage adequate generation resources, voluntary trading institutions and markets, and a plethora of service offerings, including risk management tools for all types of customers. Had California considered these factors rather than rushing headlong into adopting retail choice for the sake of retail choice, the problems being experienced today could have been reduced or possibly eliminated.

References

1. Staff Report to the Federal Energy Regulatory Commission on Western Markets and the Causes of the Summer 2000 Price Abnormalities, Part 1 of Staff Report on U.S. Bulk Power Markets, Washington, D.C., Federal Energy Regulatory Commission, November 1, 2000.
2. California's New Electricity Market, The Basics: How the California Power Exchange Works, Los Angeles, California Power Exchange, 1999: 1.
3. Ostrover, S.A., Gridlock: An enlightened look at the results of deregulation, *Comstock's Business Magazine*, November 2000: 12.

chapter four

Power marketers in a restructured power industry

4.1 Introduction

In their simplest form, power marketers buy and sell power just as a utility would. Unlike a utility, power marketers do not generally own generation, transmission, or distribution facilities, but rely on others to physically deliver the products sold. Power marketers also offer a wide variety of other services, such as risk management and tolling services, and act as middlemen for both buyers and sellers of power. The purpose of this chapter is to provide an understanding of what power marketers and their markets and services are, as well as look at recent events and the near future of the power-marketing industry.

4.2 What is a power marketer?

Power marketing refers to wholesale and retail transactions of electric power by companies other than the regulated utilities that own the distribution lines. Power marketers may buy from utility and nonutility generators, as well as other power marketers, and at the wholesale level may sell to private and public utilities, other marketers, and resellers. At the retail level, they may sell to industrial, commercial, residential, and governmental end users. Some key features of power marketers are:

1. They take title to the power being transacted, thus they assume price and market risk.
2. They must register with the Federal Energy Regulatory Commission (FERC).
3. They are not subject to state regulation.*

* In some pilot programs in New Hampshire and Massachusetts, power marketers were required to meet some state requirements to participate in the programs.

4. Their objective is to take advantage of inefficiencies in the electric power supply system through buying low, selling high, and profiting from the margin between buying and selling prices.

The Edison Electric Institute (EEI) defines seven types of power marketers:

1. Energy consultants whose primary business is to advise industries and other end users on energy and utility matters
2. Entrepreneurial firms formed to take advantage of opportunities in marketing electricity
3. Financial intermediary firms originally formed to handle financial transactions, but have branched into power marketing
4. Independent power producers (IPPs), which are nonutility entities that own generating facilities and have formed separate business units to market power from these facilities, as well as other sources
5. Large industrial firms that also engage in power marketing
6. Natural gas or other fuel-marketing firms with operations that also buy and sell electric power
7. Unregulated subsidiaries of companies with regulated utility subsidiaries

4.3 *Opening the door to power marketers*

Beginning in the late 1970s, changes in the conditions of both the marketplace and the regulatory landscape began to erode the control of the utility-dominated power industry and opened the door for the power-marketing industry. Some of the changes include:

1. Technical advances such as the gas-fired combined cycle power plants, which are more efficient and less costly than coal-fired plants, as well as advances in electricity transmission equipment.
2. The Public Utility Regulatory Policies Act of 1978 (PURPA), which requires utilities to pay avoided costs to two groups of nonutilities, (1) the small power producers using renewable sources and (2) co-generators, who sequentially or simultaneously produce electric energy and another form of energy, either heat or steam, using the same fuel source. For a small power producer to meet qualifying facility status (QF) under PURPA, it must have less than 50% ownership by electric utilities and must have at least 75% of its energy input in the form of renewable energy.
3. The Energy Policy Act of 1992 (EPAct), which substantially reforms the Public Utility Holding Company Act of 1935 (PUHCA) and simplifies nonutility generators to enter the wholesale market for electricity. EPAct allows nonutility generators, through the creation of the exempt wholesale generator category, an exemption from the PUHCA constraint that allowed holding companies to only engage

in business that is essential and appropriate for the operation of a single integrated utility. EPAct also contains transmission provisions that have led to a nationwide open-access electric power transmission grid for wholesale transactions. Anyone selling power at wholesale, including power marketers, gains the ability to seek orders from FERC that require utilities who own transmission to provide service at "just and reasonable" rates, as defined by the FERC. EPAct also gives FERC broad authority to order transmission-owning utilities to wheel, or move, power for wholesale power transactions.

4. The 1994 establishment of a "comparability standard" stating that transmission-owning utilities should offer other transmission users access to their transmission systems under the same conditions as their own use of the systems
5. The Mega-NOPR, released in 1995 by FERC, which had two goals: (1) facilitate the development of bulk power markets by ensuring that wholesale purchasers and sellers of electric energy can reach each other by eliminating anticompetitive practices in transmission services, and (2) address the transmission costs associated with the development of competitive wholesale markets
6. Order 888, released in 1996 by FERC, which (1) serves to eliminate anti-competitive practices and undue discrimination in transmission services through a universally applied, open-access tariff system in which all terms and specifications for system use are filed with FERC, and (2) ensures the recovery of stranded costs accrued by utilities in the transition to competitive markets. FERC also issued Order 889, which requires transmission facilities to electronically post information about their available capacities.
7. FERC's approval of the use of market-based rates as opposed to traditional cost-plus pricing, which led to the creation of power exchanges

4.4 Who are power marketers?

As of September 2001, there were 497 independent power marketers and 167 affiliated power marketers registered with FERC. Table 4.1 lists the purchases and sales made by both types of marketer in 2000 by quarter.

As shown in Table 4.2, for each of the four quarters, Enron Power Marketing topped the list with more than 100 million megawatt hours (MWhs) purchased. This table is a selection of major players that repeatedly follow Enron in the top ten in terms of quarterly purchased volumes. These eight major players represent more than 63% of the total purchases in 2000.*

Table 4.3 shows the number of customers, revenue, and sales for power marketers in 1999. Eighty-one percent of power marketers' customers are residential; however, these customers represent only 5.5% of sales. More than 50%

* This list is a selection of those companies in 2000 with the largest amounts of purchases.

Table 4.1 Quarterly Wholesale Transaction Totals (MWh) — Year 2000

	Affiliated		Independent	
Quarter	Purchases	Sales	Purchases	Sales
1st	454,173,775	534,825,618	105,555,344	106,082,099
2nd	434,210,246	315,456,967	114,635,011	113,177,930
3rd	689,444,036	676,160,264	169,073,412	178,662,180
4th	547,316,401	380,466,531	157,531,588	167,920,301
Total	2,125,144,458	1,906,909,380	546,795,355	565,842,510

Source: FERC Power Marketer Data, 2001.

Table 4.2 Selected Major Players' Purchase Totals (MWh) by Quarter

	1st	2nd	3rd	4th
Enron Power Marketing Inc.	101,720,868	123,497,665	165,311,100	178,153,970
PG&E Energy Trading Power	54,465,615	51,330,085	77,100,425	N.A.
Duke Energy Trading & Marketing	44,869,326	52,257,180	76,927,753	68,838,839
Reliant Energy Services Inc.	26,816,776	31,299,980	59,650,222	63,359,529
Southern Co. Energy Marketing	50,125,563	42,104,078	50,882,658	N.A.
Aquila Energy Marketing Corp.	44,698,400	39,975,298	43,269,139	58,862,924
El Paso Merchant Energy	23,523,064	22,536,418	40,375,866	26,374,825
Entergy Power Marketing Corp.	13,505,487	13,500,222	27,024,656	23,972,078
Total	359,725,099	376,500,926	540,541,819	419,562,165

Source: FERC Power Marketer Data, 2001.

of sales is to the industrial sector, with the remaining 41% in the commercial sector. Residential customers are charged a higher price than industrial and commercial customers, as shown by the average revenue per kilowatt-hour.

4.5 The power markets

Power marketers operate in two types of markets: the real-time, or spot, market and the forward market.

The spot market is a "natural market,"[1] where buyers and sellers bid on or negotiate prices in expectation of taking delivery of the product.[2] In its simplest form, this market resembles a trip to the grocery store to purchase a package of rice for a given price at that given time, assuming that the price of rice changes over time. The downside of this type of market is that it does not allow for planning, as price cannot be predicted.[1]

Table 4.3 Number of Ultimate Consumers, Revenue, Sales, and Average Revenue per Kilowatt-Hour for Power Marketers, 1999

	Number of Consumers	Revenue (000 $)	Sales (thousand kWh)	Average Revenue per kWh (cents)
Residential	566,181	170,147	4,162,053	$4.09
Commercial	109,827	1,187,693	31,394,777	$3.78
Industrial	25,361	1,299,595	40,433,571	$3.21
All Sectors	702,420	2,664,184	76,188,042	$3.50

Source: Electric Sales and Revenue, Energy Information Administration.

The forward market, often referred to as the "futures" or "contract" market, is a response to the need for planning of business activities. Forward and futures contracts also allow for varying amounts of flexibility that are not present in spot transactions.[1]

A forward contract is an agreement for the delivery of a commodity in the future at a price determined at the inception of the agreement. Terms may extend from the next day to years ahead. Forward contracts can be risky; for example, if a marketer agrees to deliver 100,000 MWh of electricity over a 1-year period at a fixed price of $20/MWh, and the actual cost of obtaining and delivering power is $30/MWh, the marketer will lose $10/MWh for a total loss of $1 million. In the absence of hedging mechanisms, discussed later in this book, a marketer will only offer forward contracts to the customer at relatively high prices.[3]

A future is a standardized contract where all terms, including delivery date, location, quality, and quantity, have been predetermined and standardized. Price is excluded from the terms and is open to negotiation. Futures are traded on exchanges such as the New York Mercantile Exchange (NYMEX)[4] and the Chicago Board of Trade (CBOT).* The exchange is responsible for reporting all transaction prices, so there is a resulting price transparency, or an ability to see, at any given time, the price at which a given future is trading. Most futures contracts are used as financial vehicles, with no intention of taking delivery of the commodity. Less than 2% of futures contracts end in delivery.[5]

There are five NYMEX electricity futures contracts that differ only in their delivery locations. Delivery locations include California–Oregon Border (COB), the Palo-Verde substation in Arizona, Cinergy, Entergy, and PJM. The CBOT has two futures contracts, one for delivery at the Commonwealth Edison hub and the other for delivery at the Tennessee Valley Authority hub.* The seller of a contract commits to deliver 736 MWh firm, or uninterruptable, electricity each month at the agreed contract price. The contract amount of 736 MWh is derived from the requirement that electricity be

* Energy Information Administration.

Table 4.4 Futures Contract "Strip"

	COB Futures Price ($/MWh)*
January	19
February	20
March	19
April	16
May	16
June	19
July	21
August	23
September	21
October	17
November	18
December	19
"Strip"	19

* Prices for illustrative purposes only.

delivered in increments of 2 MWh for 16 peak hours (there are no nonpeak futures) each business day, 23 business days a month. Finally, contracts are traded "out," or ahead of time, 18 months, which makes them useful for the creation of popular fixed-price, 1-year contracts.[3]

The yearlong contract, also referred to as a "strip," is one futures contract for each month of the contract year. When the customer requests a price quote, the power marketer calculates a weighted average of the values for each month and makes an offer to the customer. Table 4.4 shows a strip with an average price of $21/MWh for peak hours. The actual price offered to the customer will be above $21/MWh, as the marketer must adjust for overhead and profit requirements,[3] as well as risk, which will be discussed later.

The purpose of the futures market is not necessarily to act as a source of electricity, but as a financial hedge, since more than 98% of futures contracts are closed prior to delivery. Closure is achieved when the entity that purchased the futures contract, or took a "long" position, decides to sell and another party who has already sold a futures contract, or taken a "short" position, decides to buy. The alternative to the financial closing of the contract is holding the contract to maturity and the purchaser taking and the seller making physical delivery.[4]

4.6 *Services offered by power marketers*

Business acumen, coupled with the aforementioned marketplace and regulatory changes, has made it possible for power marketers to deliver a wide range of services, including facility and risk management, tolling, and customized products, as well as use a number of tools, including basis contracts, options, and swaps.

4.6.1 Risk management

Many power marketers offer risk management services, which are used to provide price stability and hedge risk. Derivatives, such as options and swaps, derived from underlying instruments, such as securities, commodities, or financial instruments, are used in the risk-management process.[3]

4.6.2 Hedging

Hedging is defined as the buying of a derivative to offset the risk of a cash position, which is the amount of commodity owned.[4] Hedging is based on physical and financial situations, both of which have "long" and "short" positions. A "long" physical position is a position in which the entity, usually either a power marketer or generator, owns the power in question. In the case of a power marketer, a long physical position would be where the marketer purchases power before finding a market. A "short" physical position is a position in which the entity, usually a power marketer or end user, does not own but has a need for power. In the case of a marketer, a short physical position would be where the marketer has sold power before securing a supply.[4]

A "long" financial position means that the entity, usually a power marketer or end user, has purchased futures; therefore, an entity with a short physical position, if they purchase futures, will have a long financial position. Conversely, a "short" financial position means that the entity, usually a power marketer or generator, has sold futures; therefore, an entity with a long physical position, if they sell futures, will have a short financial position.[4]

The key to using hedging for risk mitigation is taking opposite positions in the physical and financial situations. For instance, an entity with a short physical position needs power. An increase in the spot price will decrease their profits, as their costs are higher, and a decrease in the spot price will increase their profits as they realize cost savings. If this entity purchases futures and takes on a long financial position, the same increase in spot price will increase their profits, as the futures are worth more, and the same decrease in spot price will decrease their profits, as the futures are worth less. The converse is true for an entity with a long physical position. The countervailing effects of financial and physical positions in terms of spot price increases and decreases will assure that gains and losses are minimized, and in the case of a perfect hedge, total zero.[4]

4.6.3 Basis contracts

Basis contracts are designed to hedge against fluctuations in the price difference between two location points. Typically, one point is a NYMEX or other futures contract trading point, either COB or Palo-Verde, and the other point is a heavily traded subregion. The basis price is determined by looking at nonfirm, or interruptible, transmission rates, transmission interruptions on various routes, and the cost of electricity at the secondary point and determining the difference between the price at the contract trading point and the subregion. As is the case with futures prices, basis prices will change from month to month, although quotes for months in the same season will be similar.[3]

4.6.4 *Options*

Options give the purchaser the right, but not the obligation, to buy or sell electricity at a set price. The purchaser pays a premium, or fee, in this case in dollars per MWh, for this right. Power marketers are in a unique position to offer these services, as many utilities cannot operate quickly enough to reap the benefits of options.[6] There are two types of options: a call, or cap, option and a put, or floor, option.

4.6.4.1 *Put option*

The buyer of a put option pays a premium for the right, but not the obligation, to sell electricity at a specified price, also called the "strike" price, at a specified point in time. A simple example of the utilization of a put option is the case of a generator attempting to avoid the risk of low prices. Suppose the futures contract price is $19/MWh* and the generator would like to receive at least that amount. The generator would then buy a put option, with a $1/MWh premium paid up front. If the spot price rises above $19/MWh, the generator will sell electricity into the spot market and receive the higher price. If the price falls below $19/MWh, the generator will either (1) sell electricity to the option holder for $19/MWh or (2) sell his option at its exercise value, $19/MWh, on or before its expiration date.[4]

4.6.4.2 *Call option*

The buyer of a call option looks to avoid the risk of higher prices and purchases the right, but not the obligation, to buy electricity at a specified price and point in time. The use of call options is similar to that of put options, but instead of a generator, the simple example focuses on the end user who buys power. Suppose the futures contract price is $20/MWh, and that is the most the end user is willing to pay. The end user would then purchase a call option, with a 50 cents/MWh premium paid up front. If the spot price drops below $20/MWh, the end user will purchase electricity on the spot market. If the spot price rises above $20/MWh, the end user will (1) buy electricity from the option holder for $20/MWh or (2) sell the option, on or before its expiration date, for its exercise value of $20/MWh.

4.6.4.3 *No-cost collar*

A no-cost collar establishes upper and lower price limits through the use of put and call options. The collar is "no-cost" because the premiums spent on

* The futures contract and premium prices used in the examples are for illustrative purposes only.

the purchase of the put option will be cancelled out by the premiums gained through the sale of a call option to another party. For example, a power producer purchases a put option, at a $1/MWh premium with a strike price of $20/MWh, because, as was the case in the previous put example, the producer does not wish to sell electricity for less than that amount. At the same time, the producer sells a call option to another party, at a $1/MWh premium with a ceiling price of $25/MWh, as the other party does not wish to spend more than $25/MWh for electricity. The premiums cancel out and the power producer has the security of knowing in what range its electricity will be priced.

4.6.4.4 Price swaps

A price swap is a negotiated agreement, sold over the counter and not on an exchange, between two parties to exchange, or swap, specific price-risk exposures over a predetermined period of time.[4] In its simplest form, a power marketer or customer will exchange a fluctuating rate for a fixed rate based on the Dow Jones Telerate or McGraw-Hill Power Markets Week indices.[3]

For example, Customer A is paying a fluctuating rate for electricity, which is currently at 2.5 cents/kWh.[3] In an attempt to create a budget, Customer A desires a fixed price for electricity, so he buys a swap with Customer B, who is the counterparty in the transaction. Customer B has a fixed rate of 4.5 cents/kWh and wants to lower his electricity costs, even though swapping for a fluctuating rate involves risk that electricity costs will rise. The transaction will break down as follows, assuming Customer A and Customer B use the same amount of electricity:

1. Customer A will pay a fixed amount of 3 cents/kWh to Customer B. Because Customer A's goal is price stabilization rather than profit, he willingly pays a half-cent/kWh premium over the 2.5 cents/kWh fluctuating rate; 3 cents/kWh becomes Customer A's final rate.
2. Customer B pays 2.5 cents/kWh to Customer A's electricity provider, in effect paying Customer A's electric bill. If the rate were to change, Customer B would pay the increase or decrease.
3. Customer B also pays a fixed 1.5 cents/kWh plus the 3 cents/kWh from Customer A to his own provider to cover the entire fixed 4.5 cents/kWh cost.
4. Customer B's final electricity rate is 2.5 cents/kWh, paid to Customer A's provider, plus the fixed 1.5 cents/kWh to his own provider for a sum of 4 cents/kWh, a half-cent/kWh savings over his previous situation.

Unfortunately, most of the time it is not possible to find two perfectly matching customers such as Customer A and Customer B. This is where power marketers come into play. A power marketer who deals in swaps is called a "market maker," and the market maker establishes standardized prices and price indices and can handle many transactions quickly and easily. This allows swap transactions to be free of unhedged leftovers, or "nubs."[3]

4.6.5 *Facilities management*

Facilities management services work to improve the utilization, maintenance, and operation of a customer's existing plants, personnel, and other assets. These services are usually provided in addition to more traditional services, such as the selling of a customer's excess power production. The savings that result from facilities management services are usually shared between the power marketer and the customer under terms agreed upon before the service is rendered. There is a significant risk benefit to the customer using a power marketer to provide facilities management service, as oftentimes the power marketer will assume the up-front costs and risk of the project. The cost of the transfer of risk is usually a higher percentage of the resulting savings going to the power marketer.[7]

4.6.6 *Total energy services*

Total energy services is the cost-reducing process of combining many types of input fuels, such as coal, natural gas, and oil, and end-use energy or services, such as electricity, steam, and hot water, and switching them around when a price decrease would result. The power marketer thus becomes the fuel procurer, which entails the assumption and management of risk, as well as a sharing in the resulting cost savings.[7]

4.6.7 *Tolling services*

Tolling is the process in which the power marketer buys raw energy at low prices at one location and time and sells a portfolio of energy services and customized products to a customer at another location and time. This process, combined with the aforementioned risk management tools, can reduce risk to the individual customers.[7]

4.7 *Conclusion*

Power marketers do more than just buy and sell power; power marketing is an industry growing in both function and importance. Power marketers perform a wide variety of tasks in order to provide energy and financial services to their clients in the wholesale market and, increasingly, in the retail market. Future changes in regulatory and marketplace structure will only serve to increase these roles. These power marketers play an important role in competitive markets with their ability to reduce transactions and information costs, thereby linking a greater number of buyers and sellers. The arbitrage and quick profit-taking nature of this business often leaves it with a less-than-positive reputation. The California crisis is a case in point.

Nevertheless, these marketers are vital to equate markets across time and geography. It is the high profit potential that gives them the incentive to participate and to take the high risks to equilibrate the market.

References

1. Hossein, R., *Fundamentals of Petroleum Trading*, New York, Praeger, 1991: 4.
2. Johnston, D., *Oil Company Financial Analysis in Nontechnical Language*, Tulsa, PennWell Books, 1992: 314.
3. Spiewak, S., *Power Marketing: Price Creation in Electricity Markets*, Arlington, The Power Marketing Association, 1998: 4.
4. Stoft, S., Belden, T., Goldman, C., and Pickle, S., *Primer on Electricity Futures and Other Derivatives*, Lawrence Berkeley National Laboratory, Berkeley, 1998: 13.
5. Treat, J.E., *Energy Futures: Trading Opportunities for the 1980s*, Tulsa, PennWell Books, 1984: 4.
6. Spiewak, S., *Uses of Options in Electric Power Marketing*, Arlington, The Power Marketing Association, 1998.
7. Sioshansi, P. and Altman, A., Power marketers: Who are they and what do they do? *Electricity Journal*, December 1998.

chapter five

The role of distributed energy resources in a restructured power industry

5.1 Introduction

One of the more interesting technological innovations during the past several years has been associated with distributed energy resources (DER). DERs, simply put, are small power generation and storage applications, usually located at or very near customer loads. The application of these small-scale power technologies is gaining widespread interest and acceptance due to their ability to further customer choice and competition. Locating power generation and storage technologies allows customers to balance their cost/reliability preferences in ways that were previously very limited. The prime benefits of DER, however, are associated with their interconnected nature with the utility distribution company (UDC) grid. Under a properly structured environment, these benefits can run in two directions: one for the customer, and the other for the UDC.

The flexibility, size, cost, and modularity of DER create significant benefits. These benefits include:

1. Reliability: DER can provide on-site backup close to customer loads. High reliability is becoming increasingly important for high technology and digital applications that are sensitive to outages.
2. Power Quality: Voltage sags and surges can damage digital equipment, including computers, Internet servers, and telecommunications equipment. Many DER technologies can deliver high power quality, but in many instances, there has to be balance with cost.
3. Energy Efficiency: DER can be used to customize usage profiles for peak shaving applications. For some larger uses, combined heat and power (CHP) applications further on-site energy efficiency opportunities.

4. Cost: DER, particularly those applications that facilitate older, rebuilt equipment, can be used to provide very low-cost, on-site power needs. For UDCs, DER can be used to avoid or, in some instances, defer, costly distribution level upgrades.

Promoting DER applications that are interconnected to the distribution grid is important and cannot be stressed enough. Stand-alone applications significantly reduce the number of benefits to both UDCs and end users. There are a number of financial disincentives that UDCs will have in supporting any customer-initiated DER applications. Reducing or eliminating these disincentives will be tantamount to DER success. This chapter addresses the competitive opportunities, as well as the current barriers, to DER applications.

5.2 A definition of DER

While DER refers to a broad range of technologies and applications, most attention is being directed at opportunities to self-generate electricity. The four major distributed generation technology categories include: reciprocating engines, gas turbines, microturbines, and fuel cells. Table 5.1 presents the cost and operating characteristics of a number of commercially available or developing DER technologies.

Currently, reciprocating engines or simple combustion engines are some of the older and more mature of the DER technologies on the market. These prime movers can range from newly developed, high-efficiency engines to rebuilt older engines used for transportation purposes. Gas turbines are another popular on-site power generation technology that offers a range of readily available applications. Like reciprocating engines, these turbines can be either newly developed, high-efficiency turbines or rebuilt turbines previously used in aviation applications.

The popularity of reciprocating engines and gas turbines is based upon availability, low capital cost, modest exhaust emissions, extended service intervals, long service lives, and well-developed sales and marketing infrastructure. The installed capital costs of these technologies range from $150/kilowatt (kW) for a 200-kW unit to $724/kW for a 12.5-kW unit. Gas turbine costs are equally competitive and run between $350 to $1000/kW.

Microturbines and fuel cells are two emerging technologies that are attempting to challenge the reciprocating engine market. Microturbines are essentially mini-jet aircraft engines that in many ways are based upon the same aerospace technologies that revolutionized the larger-combustion turbine market of the electric power industry during the past 10 years. These technologies are much smaller than their larger aviation-based counterparts and offer many size advantages that larger turbines are unable to deliver.

Fuel cells, on the other hand, facilitate a chemical process that acts like a big battery, which makes rather than stores electricity. Unlike many energy-conversion processes, fuel cells generate electricity in a continuous, direct process, thereby reducing the excessive energy losses prevalent in other

Table 5. 1 Examples of DER Costs under Differing Technologies

Cost and Operating Performance Categories	Fuel Cell	Microturbine	Microturbine/CHP	Reciprocating Engines	Reciprocating/CHP
Capital costs ($/kW)	2000	800	800	450	450
Capacity (kW)	200	400	400	400	400
Capacity Factor	0.95	0.95	0.95	0.95	0.95
Net Annual Generation (kWh)	16,644,000	3,328,800	3,328,800	33,288,000	33,288,000
Total Capital Cost ($)	400,000	320,000	320,000	180,000	180,000
Finance Costs ($)	40,000	32,000	32,000	18,000	18,000
Capital Costs ($/kWh)	0.2644	0.1057	0.1057	0.0595	0.0595
O&M ($/kWh)	0.05	0.005	0.005	0.005	0.005
Heat Rate (Btu/kWh)	6000	10,000	8000	13,000	10,000
Fuel Costs ($/MCF)	2.25	2.25	2.25	2.25	2.25
Gas Use (MMBtu)	9986	33,288	26,630	43,274	33,288
Total Fuel Cost ($)	22,469	74,898	59,918	97,367	74,898
Fuel Costs ($/kWh)	0.0135	0.0225	0.0180	0.0293	0.0225
Estimated Levelized Cost	0.2829	0.1332	0.1287	0.0937	0.0870
Intererst (Annual Percent)	0.1000	0.0800	0.0800	0.0800	0.0800

Source: From Priddy, R.D. and Dismukes, D.E., *Distributed Energy Resources: A Practial Guide for Service*, Ft Energy, Boulder, CO, 2000: 74. With permission.

multistep energy conversion applications. These applications are particularly useful for those end uses that require high on-site power quality.

There are a number of small power application advantages that microturbines and fuel cells have over conventional reciprocating engines that make them very attractive. These technologies tend to have lower emissions and noise levels, excellent combined heat and power applications (i.e., cogeneration), somewhat better efficiencies, and provide power stability capabilities. Unfortunately, both microturbines and fuel cells are still very expensive relative to their main DER competitor: reciprocating engines. Fuel cells, for instance, are currently priced around $1500/kW, a figure comparable with solar (photovoltaic) energy resources. Microturbines, while somewhat lower in costs, are still expensively priced at $1000/kW. These high installed costs keep both of these technologies from being fast market movers in the very near future.

A benefit of DER applications is that these various technologies enable customers to choose their own levels of reliability and/or cost. Today retail customers are becoming increasingly more sophisticated in their power needs and requirements. A number could benefit from moving away from "plain-vanilla" service as has been traditionally provided by utilities. In examining these trade-offs, it is important to consider that in the past, the degree of "economic" and "reliable" power has been determined in large part by regulation. Even today, as more retail markets become increasingly competitive, historic traditions in service provision remain, particularly when it comes to customers taking service under default, or "provider of last resort" terms.

In addition to the reliability–cost trade-off discussed above, a number of other types of applications and trade-offs can be considered by different DER providers and customers. The perceived advantages and applications among all of these potential stakeholder groups are reasons why DER has come into vogue over the past several years. Table 5.2 presents the different applications and stakeholder groups that can be impacted by DER — and the advantages of these applications to the various groups.

Despite these numerous opportunities, there are, and continue to be, a number of barriers to DER applications, many of which are similar to the barriers faced by industrial cogeneration applications allowed under the Public Utilities Regulatory Policy Act (PURPA) in the early 1980s. Several barriers are the result of traditional regulation and how the regulated retail restructuring process is unfolding in different states. Understanding these disincentives is important to understanding why these technologies have had trouble securing a strong foothold in today's energy markets and may offer some insights into future challenges.

5.3 *UDC disincentives associated with developing DER applications*

Despite industry restructuring, regulation will continue to play an important role in the delivery of electricity to end users. Earlier chapters, for instance,

Table 5.2 Stakeholder Benefits Associated with DER

Stakeholder Group	Combined Heat and Power	Standby Power	Peak-Shaving	Grid Support	Stand-Alone
Customer	Lower energy costs, higher overall reliability	Avoid economic loss due to system outage and satisfy critical support systems	Lower peak-period energy costs	Customers generally benefit from the enhanced service provided, but may be isolated from competition markets as a result	Customer option to avoid high-cost backup service, remote communications, and control systems
T and D System	Positive to negative, depending upon the application	Can be integrated with utility needs to provide both customer and grid benefits	Can be integrated with utility needs to provide both customer and grid benefits	Enhances grid stability and economic customer service	Loss of customer load and associated revenues
Energy Service Provider	Power and heat can be separately marketed; ESPs can also provide ancillary services to CHP customers	Can facilitate ESP marketing of interruptible power supplies; widely used strategy of municipal systems	Can aggregate and sell customer peak-period generation	Possible benefits as an owner/operator of the system	Possible benefits as an owner/operator of the system
Natural Gas Industry	Benefit from high gas consumption, possible fuel switching benefit for oil-fired boilers	Minimal impact, but cost to service customers is high	Good match of gas off-peak period with electric on-peak period	Generally similar to peak-shaving benefits	Benefit from high gas consumption
Society	Environmental benefits with some technologies, energy efficiency, economic development	Public health and safety	Environmental and energy-efficiency benefits	Environmental and energy-efficiency benefits	Less likely in a competitive market to represent an optimum allocation of resources

Source: From Priddy, R.D. and Dismukes, D.E., *Distributed Energy Resources: A Practial Guide for Service*, Ft Energy, Boulder, CO, 2000: 74. With permission.

noted that distribution system operations and rates will continue to be regulated. The impact of regulation on distribution, and as a consequence on DER, cannot be emphasized enough. As noted earlier, DER resources create the largest value when they are interconnected into the distribution system. The benefits of DER can deteriorate rapidly for nongrid, stand-alone or "cut–the-wires" applications. However, attempting to maximize the opportunities for cost-effective DER is difficult, since many distribution utilities could face a number of serious disincentives for supporting DER. The disincentives for utilities can be strong, since the loss of sales on every kilowatt-hour (kWh) of self-generated electricity is a kWh of sales that has been lost by the utility or its affiliate.

Three of the more common barriers for DER are related to standards for interconnection, pricing, and backup service that are influenced heavily, if not completely, by regulation and the continued monopoly status of distribution operations and services in restructured markets. The most pervasive barriers that are arising as more states move to retail competition are associated with physical interconnection terms for DER applications, rate design, distribution level wheeling, and even stranded costs. The following sections consider each of these fundamental disincentives and their implications for DER implementation.

5.4 Interconnection issues

One of the major barriers associated with DER implementation is the inability to easily and seamlessly interconnect with the utility distribution grid. This interconnection usually occurs at the distribution level. For most small-scale applications, interconnecting a distributed resource at the distribution level would entail hooking the application to a three-phase distribution line at the 69-kilovolt (kV) level. Interconnection terms for DER are typically defined by UDCs and, in some cases, their regulators. There are a number of similarities between today's DER applications and its large-scale industrial cogeneration counterparts that were promoted through PURPA. The significant difference, however, is that no federally mandated requirements for DER facilitate interconnection, emergency and standby power, and buyback rates for excess on-site power. In states where regulators have not required utilities to develop favorable DER rules, the challenges are even more difficult, since interconnection, at this point, is directed by the liberties of the host UDC.

If these applications are discouraged through the interconnection process, there will be a socially inefficient level of DER developed. Some of the more significant barriers to DER are associated with the initial interconnection process that includes initial request and terms for projects, interconnection studies and fees, and any required facilities upgrade costs. Some states have tried to reduce these barriers through a number of streamlined rules to lower informational and administrative costs for DER applications.

California, for instance, established Modified Rule 21, which was designed to streamline the interconnection of new, small-scale generating facilities, thereby relieving California's electricity supply constraints and encouraging self-generation while maintaining a comprehensive and user-friendly application form. To achieve this objective, standardized and simplified utility interconnection protocols and standard tariff language was developed. Rule 21 is part of each investor-owned utility's tariff.

This rule requires the applicant to supply sufficient information to allow the utility to accurately evaluate the interconnection requirements for a facility, yet it is not supposed to be so burdensome as to serve as a barrier to entry. Some of the other highlights of the California interconnection rule include:

1. The total cost of the initial and supplemental review not exceeding $1400.
2. Continuing to update a distributed generation (DG) interconnection database. This would provide a readily accessible inventory of distributed generators in California, regardless of size.
3. A limitation of liability clause and increased minimum insurance requirements that differ based on size and customer class.
4. The adoption and monitoring of the interconnection application form through a post-implementation working group to determine if additional changes are required.

5.5 *Rate design issues*

While interconnection is both a physically and economically important issue, rate design can also have significant implications on DER applications. These rates will be important in sending price signals to potential DER projects. If these rates are developed in a relatively inefficient manner, a less-than-optimal level of DER will be developed. An optimal rate design would be a distribution delivery rate, or family of rates, that varied by time (system loading), location, desired firmness of capacity, and volume (kWh).

Ultimately, finely differentiated options in rate design would allow all customers of the distribution system to select the type of service each desires and influence their own service costs. When the time dimension, reflecting system loading, is finely differentiated, then the need for standby, maintenance, emergency, or curtailable tariffs is moot. Customers will pay for the delivery service depending upon when it is used, where it is used, how firm its delivery assurance, and how much is needed. The reality of setting distribution rates, particularly in a regulated environment, tempers the opportunities for developing optimal rates. As a consequence, second-best solutions will need to be developed that include the introduction of other tariffs, such as standby and curtailable, to account for the ways that customers might group themselves together.

The ongoing rate design practices of most regulatory commissions, however, will continue to pursue a number of other equally important policy considerations that include the goals of keeping rates fair, affordable, and understandable for less-advantaged customers. Universal service is an important motivation for these policy goals. Because of these "realities" of regulatory rate making, it seems likely that most DER applications will have to settle for discrete, yet differentiated, distribution rates that take into consideration the nature and costs associated with the various types of distribution service.

Another consideration in the distribution-level rate-making process is the balance between fixed and variable charges for service. Traditionally, utilities have created rate structures that are commonly referred to as "two-part tariffs." Such a structure levies a fixed customer charge and an incremental, kWh-based charge for service. As noted in earlier chapters, the unbundled process will attempt to separate these two-part tariffs into generation, transmission, and distribution rates. However, during the course of this process, the regulators are being faced with proposals to move more, if not all, of the distribution-oriented revenue requirements into fixed rates.

The justification for basing an increasing proportion of its revenues into fixed charges is founded upon a relatively static short-run approach to pricing and costs. UDCs argue that their distribution system is relatively static and based for peak-day capacity. As a result, variable costs of running the system (in the short run) are relatively small, hence the justification for recovery of the revenues from customer charges and fixed-tariff rates.

What this philosophy overlooks, however, is the long-run nature of distribution costs, the incentives for future infrastructure investment, and the benefits that DER could provide in deferring or changing a number of these investments. As noted earlier, optimal rates for distribution service should reflect the long-run incremental costs of providing these services. The long-run costs should include the capacity additions and capital upgrades required to provide distribution service. These additions should figure into distribution rates and send signals to the market about when and where DER is the most economical.

Standby and backup power is a critically related rate issue that impacts the future development and opportunities for DER. Most recent proposals for DER standby service that are being offered by UDCs do not accurately reflect the probability of failure for on-site generation resources and essentially price DER on comparable terms with full-requirements customers. Setting grid charges to end users that are equal to either the customer peak demand or the total capacity of on-site generation essentially prices the risk of equipment failure at the system peak cost. In other words, such an approach assumes that DER applications have a 100% probability of failure at peak levels. Such a pricing approach eliminates using DER as a peak-shaving application and essentially limits on-site generation applications to baseload or nothing. Such approaches substantially limit the possibilities for DER.

Appropriately developed standby charges should reflect the DER customer's choices regarding the level of firm vs. nonfirm standby service, the quantity of standby capacity desired, the location of the customer, and the probability of using standby capacity during peak constrained times vs. using standby capacity at other times. Regarding the last element, if the standby rate does not change as a function of time (system loading), then an estimate of the likely pattern of loading will need to be developed in order to appropriately structure the standby rate. For instance, one would expect to see higher standby rates during peak periods in relatively congested, dense urban areas than during off-peak periods in relatively less congested areas.

In terms of firmness, standby service could be differentiated through various degrees of reserved distribution capacity. For instance, customers should be allowed to enter into agreements with UDCs for distribution service based on different levels of firm vs. nonfirm capacity. Firm distribution capacity reservations should command a higher price (or premium) than nonfirm reservations. This type of service differentiation facilitates customer choice because it allows the customer, and not the UDC, to select the appropriate degree of distribution reliability.

In terms of volume, standby rates should not be set at levels that reflect the maximum peak demand at a particular load location nor the maximum level of installed DER capacity at a particular location as though its utilization of the distribution system during peak contract time was 100 %. Standby service needs to reflect the probability of providing emergency power during constrained times vs. during nonconstrained times.

However, standby rates set at total installed DER capacity, and at prices similar to full-requirements customers, assume that standby service will be required for a simultaneous failure of all on-site generators. This is a low-probability event that would rarely, if ever, happen. Setting rates based on this contingency forces DER customers to pay for capacity they do not want and would not use. Clearly, this is not a rate design mechanism that helps facilitate customer choice.

In reality, on a system of distributed generators, the likelihood of more than 10% of those generators being unavailable at any point in time is very low. Therefore, the line capacity needed to meet the standby requirements of these distributed generators should be about 10% of the total generating capacity of the distributed generators.

5.6 *Wheeling power at the distribution level*

Distribution wheeling could be facilitated by regulatory rules and recognition of the immediate opportunities for DER wheeling. Most opportunities for distribution wheeling do not envision selling power on the bulk market in competition with merchant power facilities. Likewise, most applications do not envision wheeling power over long distances over expansive distribution networks. The more immediate opportunities for low-voltage-level wheeling are primarily within a very tight geographic locality.

The primary opportunity that many DER applications would like to facilitate is one where power could be transmitted from one location to another noncontiguous location in order to maximize this customer's energy-choice options. Consider, as an example, a hypothetical manufacturer who has his primary operations at one location, his recycling facilities in a warehouse at a different location across the street, and storage and packaging operations two blocks down the street. All of these operations are on the same feeder, but they are not contiguous.

This manufacturer has what appears to him to be a simple goal: he would like to have a DER application at the main manufacturing location and size the unit large enough to supply all three sites. In addition, he would like to take advantage of a cogeneration opportunity at the manufacturing site that does not exist at the other two locations. If the manufacturer could wheel his power to his nearby affiliates, he could maximize his total energy opportunities.

5.7 Conclusions

DER technology and economics are improving rapidly and being promoted by an increasing number of credible and committed manufacturers. While many of the technologies are still in early stages of commercial implementation, a number of energy service companies are recognizing the opportunities to increase customer value through on-site generation. The next level of distribution-level planners will need to put these opportunities, and their challenges, into perspective.

Despite the opportunities in the energy services business, regulation will still be a pervasive theme conditioning this marketplace. Since DER applications will be tied to the distribution end of the electric power business, regulation will be as important as the economic and technological aspects of these technologies. As the industry evolves, there will be a challenge to balance the engineering and safety concerns of these technologies with their regulatory and economic implications. Viewing these resources as opportunities rather than threats will go a long way in balancing these different interests.

chapter six

Independent power generation

6.1 Introduction

One of the pressing challenges in today's energy industry is the development of supporting infrastructure. Nowhere is this more readily apparent than in the electric power industry. Years of upheaval, uncertainty, and regulatory change have clearly had consequences that are taking their toll today. What is unique about today's energy industry revival is the development of competitive, as opposed to regulated, forces for driving the nature and the direction of energy infrastructure investments.

The power generation sector, in particular, has seen a virtual explosion in announced construction activity during the past several years. This increase in industry activity is the result of a confluence of different factors, including the following:

1. Technological: Over the years, smaller, more modular, and more efficient power-generation technologies have emerged.
2. Economic: The nature of wholesale* power markets has changed from one in which pricing and market conditions were determined by regulation to one in which the market determines the amount and prices of electricity to be offered.
3. Public Policy: Transmission systems have been legally opened to support open access and nondiscriminatory transportation of power across utility power grids.
4. Institutional: New market mechanisms and institutions have arisen that facilitate the trade of bulk (wholesale) power as a commodity.

* This chapter focuses exclusively on the impact that merchant facilities have on wholesale power markets. Here, wholesale power markets are defined as bulk power markets where purchasers are not the ultimate end users of electricity. A wholesale power market transaction is one where a utility that is short on capacity purchases electricity from another utility (or merchant plant) in order to supply power to its own customers. Wholesale competition allows these trades to occur outside regulation with prices being negotiated between the two utilities. Retail markets, on the other hand, are defined as markets where the customers are the ultimate end users of the energy being purchased.

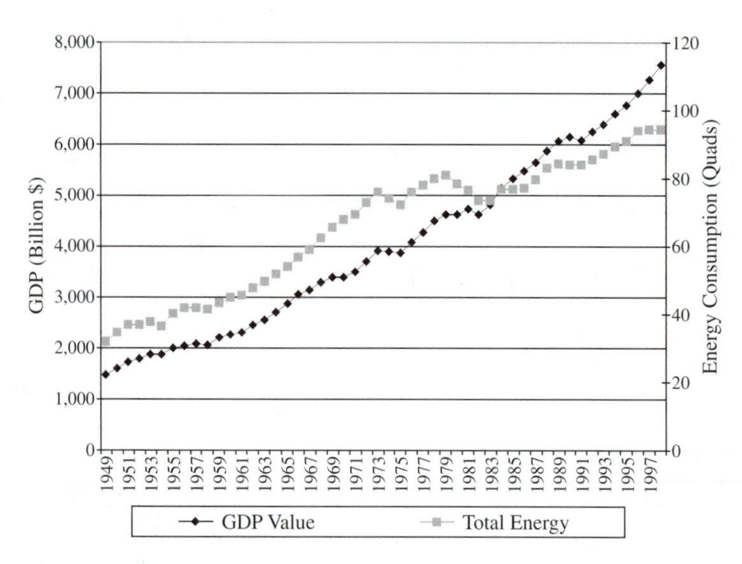

Figure 6.1 Annual total electric energy consumed and U.S. gross domestic product.

An increasingly important consideration in the energy industry is the role it plays in securing economic growth opportunities. The relationship between energy and economic growth during the past 50 years has been well established by academic literature.* Figure 6.1 shows this relationship for the U.S. economy quite clearly.

The electric power industry has transformed the relationship between energy and economic growth even further. Throughout the post-war period, the U.S. economy has undergone a dramatic transformation from one based upon primary-fuel driven, mechanical industries to one that increasingly emphasizes high technology, digital and computer applications, and increased complexity.

If economic growth is to be maintained in this increasingly more digital "new economy," additional competitive generating capacity must be developed. Businesses and households are hurt and lose real disposable income as a result of expensive and unreliable power. A recent study by the Federal Reserve Bank of San Francisco, for instance, noted that the increased energy costs** associated with the California energy crisis would set households back by $450 more per year — or 1% of the median household income. This could rise to as much as 1.5% of total household median income if businesses pass their increased cost along to consumers.

* See Jorgenson, D.R., The Role of Energy in Productivity Growth, *American Economic Review* 74(2), 1984, 26–30 for a seminal discussion on this relationship. A more contemporary article was prepared by Moroney, J.R., Energy Consumption, Capital and Real Output: A Comparison of Market and Planned Economics, *Journal of Comparative Economics*, 14(2), 1990, 199–220.

** Increased energy costs include retail natural gas and electricity costs to consumers.

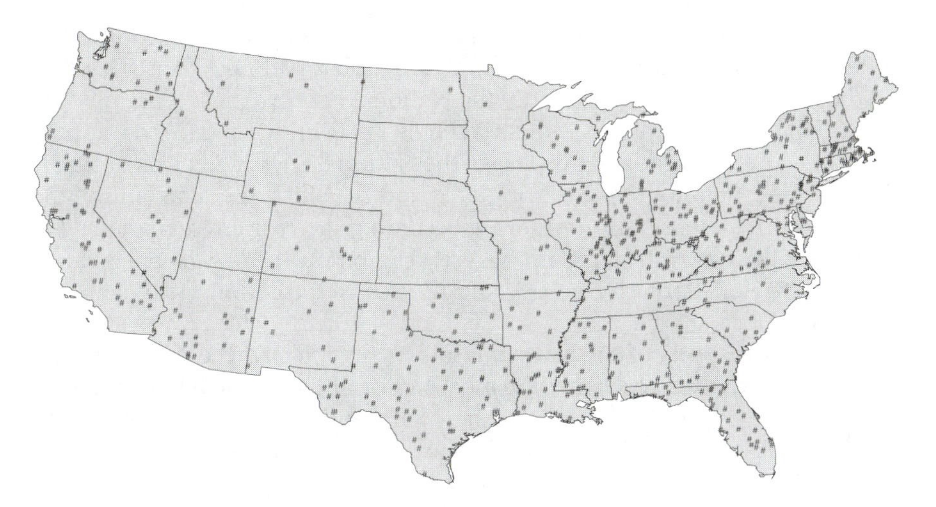

Figure 6.2 Announced independent power projects in the United States.

The San Francisco Fed also noted that these decreases in household income have been substantially lessened because of subsidized prices by the state of California. The recent study noted the following:

> If the full rise in wholesale electricity prices — much of which currently is being covered by the state as a result of the procurement of power by the Department of Water Resources — was taken into account, our estimate of the increase in energy-related expenditures by the average California household would rise substantially.[1]

Fortunately, market incentives in most regions of the United States seem to be working. Industry changes and market forces have stimulated new power plant construction activity. Today, for-profit independent power providers are constructing the next generation of power facilities; this is unlike the past when power generation facilities were built almost exclusively by regulated utilities. Figure 6.2 shows the number of independent power plant construction projects throughout the United States.

6.2 The origins of competitive wholesale markets

One important factor changing the nature of electric power markets has been the advent of competitive opportunities for new sources of power generation. Quickly fading is the past regime of regulated prices, as well as limited opportunities for trading, profits, and energy efficiency. The origins of competition, however, are not new and can be dated to the late 1970s when the energy crises changed public policy. This is when the notion came about that utilities were "natural monopolies" and should be the only regulated providers of electricity in the marketplace.

In 1978, Congress passed the National Energy Act, which comprised five different statutes: (1) the Public Utilities Regulatory Policy Act (PURPA); (2) the National Energy Tax Act; (3) the National Energy Conservation Policy Act; (4) the Power Plant and Industrial Fuels Act (PPIFA); and (5) the Natural Gas Policy Act. The general purpose of the National Energy Act was to ensure sustained economic growth during a period in which the availability and price of future energy resources were becoming increasingly uncertain. The two major themes of the legislation were: (1) promote the use of conservation and renewable/alternative energy and (2) reduce the country's dependence on foreign oil.[2]

While all aspects of the National Energy Act affected the electric power industry, PURPA was probably the most significant, because it was designed to encourage more efficient use of energy through nonutility cogeneration. The statute requires utilities to interconnect and purchase power from any qualifying facility (QF) at a rate not to exceed the utility's avoided cost of generation.* This policy, while originally designed to promote energy efficiency, had the unintended consequence of encouraging the development of a plethora of new sources of electric generation in an industry that, as conventional wisdom held, was a natural monopoly. The emergence of these nonutility generators proved that entities other than utilities could construct and operate power plants efficiently and reliably.

Since the 1980s, the power generation business continued its trek toward greater levels of competition and efficiency. By the early 1990s, Congress decided to take the unintended policy consequences of PURPA one step further by enacting the Energy Policy Act of 1992 (EPAct). The legislation is important for two reasons. First, EPAct created a whole new class of power providers called "exempt wholesale generators" (EWGs) that are essentially competitive independent power plants and not subjected to traditional rate-making regulation. Second, the EPAct allowed the Federal Energy Regulatory Commission (FERC) to require regulated electric utilities to "wheel" (transport) power across their regulated power transmission grids.

These two developments, taken together, created a new class of generation market participants, a new market for the generation of electricity, and a new means of transporting (or wheeling) electricity to these markets across the entire United States. FERC promulgated the final rules outlining the terms and conditions for the open and nondiscriminatory use of the electric power grid in 1996 in its industry-renowned Order 888.

Order 888 was instrumental in opening the wholesale power market to competition and facilitating independent, or what is commonly referred to as "merchant," power. Without Order 888, competitive power generation firms would have been able to construct and operate their facilities, but

* Avoided costs are defined as the utility's cost to produce a marginal unit of electricity. Only cogeneration facilities and renewable energy small power production facilities are entitled to these provisions. Cogeneration is defined as the combined production of thermal and electrical energy. Most cogeneration applications capture steam that previously would have been vented into the environment and use it to produce electricity; hence, the term cogeneration.

would have been required to deal directly with transmission-owning utilities for moving their power to wholesale customers. Without these rules in place, transmission-owning utilities would have been able to give preference to their own competitive (or regulated) generating facilities at the expense of their potential competitors. This new order helped create a system in which transmission lines, regardless of ownership, would serve as a common carrier to facilitate wholesale trade. From 1996 on, competitive sources of electricity have been able to compete on a level playing field with incumbent utility generation.

The promulgation of Order 888 transformed the industry. In addition to creating a competitive power market, it also helped facilitate the growing convergence between the power business and other energy industries. New trading mechanisms and institutions that arose in the aftermath of Order 888 served to facilitate this process.

Today, independent power providers play an important role in regional power markets. The nature of these providers, however, is often misunderstood. Independent — or merchant — power plants are those facilities that are usually constructed and operated by independent companies (i.e., nonutility companies) for a potential profit. These facilities, and their developers, differ in important ways from other utility and nonutility sources of power generation.

Utilities, for instance, are regulated monopolies that have a guaranteed retail customer base. Prices are set by state regulators to curb potential monopoly abuses. As monopolies, utilities are allowed to recover their prudently incurred costs, and to have the opportunity to earn a reasonable rate of return on prudently incurred capital investments. In return for their monopoly status, utilities are required to provide safe, reliable, and economic service to their customers.

Other nonutility power generating sources, primarily qualifying facilities or cogenerators under PURPA, are not in the primary business of producing electricity. These facilities typically produce some product and generate electricity as a secondary endeavor. If these types of nonutility cogenerators meet thermal and other ownership and operating requirements established by FERC, they are entitled to sell their power to utilities based upon the utilities' avoided cost. They are also entitled to emergency, standby, and backup power should their on-site generating facilities go down for planned or unplanned outages.

Competition in wholesale markets over the past several years has not come without its share of growing pains. Some of the more painful recent experiences of this process have included the following:

1. The past several summers have seen an increase in the price volatility of wholesale power markets.
2. In addition to price volatility, wholesale markets have experienced a number of incredible price increases in absolute magnitude. In some

instances, wholesale power market prices have reached levels of $10,000/megawatt-hour (MWh) on certain super peak hours.
3. The integrity of a number of "new players" in the market has been challenged. These players did not understand and did not anticipate the nature and volatility of the new environment and were caught short on their respective power purchases and sales.
4. Outages have increased, power reliability has been challenged, and capacity margins throughout a number of regions in the United States are falling because of continued strong economic growth (stimulating demand) and an apparent shortfall of existing generation resources and infrastructure.
5. Markets can be both integrated and segregated given varying conditions on the electric power transmission system. The operation of this system is important in determining access to alternative power supplies.

These recent experiences have highlighted a number of important lessons about electric power markets. First, and most important, are physical power generation matters. Despite all of the innovations in trading mechanisms, financial instruments, innovative transmission pricing regimes, and theories about power markets, the importance of having physical supplies of electricity (i.e., power plants) cannot be underestimated. Paper transactions are limited in their ability to keep the lights on. Eventually, these trades and transactions will have to be delivered. Recent events in California have shown that in the absence of physical power generation, strong demand for electricity can only be met in two ways: either prices must rise to lessen demand or demand must be curtailed through interruptions and rolling blackouts in instances where power is simply unavailable.

Second, the separation between wholesale and retail markets is artificial. Eventually, the ramifications of power purchased at the wholesale level will ripple down to retail customers — even if those customers are under traditional regulation. Today, many utilities in states that have not moved forward with retail choice are generation strapped, for a number of different reasons, and have to purchase electricity on the wholesale market. When these utilities purchase electricity on behalf of their retail customers, the costs are usually directly passed on to those customers in their monthly bills. Thus, as these wholesale purchased power costs increase, so too have residential, commercial, and industrial electricity bills.

Third, the regulatory environment can strongly influence the siting decisions of competitive independent power plants. Clearly, a correlation exists between siting decisions and a state's movement toward electric restructuring. However, this is not the only factor influencing independent power plant siting. Consider that California, for instance, was the first state in the nation to adopt electric restructuring. Over the past 17 years, the Western Systems Coordinating Council (WSCC), which encompasses the entire western portion of the United States, has been experiencing substantial growth in peak demand. The annual average growth in peak demand for California during this period (1982–1998) was approximately 3.2%, compared to an annual average increase in generating capacity of less than 1%.[3] The apparent shortfall in capacity, coupled with the new competitive retail opportunities, has thus far failed to entice a large number of independent facilities.

Equally important are other factors, such as policy stability on tax and environmental issues, that can have equally important implications for the construction and operating costs of a new multimillion-dollar power plant. California, for instance, with its stringent environmental laws, rules, and standards, is not considered by many developers as being friendly toward power-plant siting. While the state has recently changed these rules to allow "fast-track" approval processes, many of these developments will take time — hardly a concession to ratepayers suffering from high rates and poor reliability.

6.3 *Who are independent power developers?*

Independent generators, unlike regulated utilities, do not have a guaranteed retail customer base for their electrical output. These providers must market their output and, as a result, are allowed to charge market-based rates and earn market-based returns on their investments. Independent generators differ from such other nonutility sources of power as cogeneration in two important ways. First, they are not end users of electricity and do not use their electrical output on site. Second, regulated utilities are not obligated to purchase any of the competitive independent power provider's output.

Independent providers come from a variety of corporate backgrounds. A listing of the top independent power developers has been provided in Table 6.1. A number of these developers arose to take advantage of the business opportunities offered by the restructured power business. These include companies like Calpine, Cogentrix, and Panda Energy. Several others, however, are the unregulated affiliates of companies traditionally associated with utility operations. These include TECO Energy, Duke, and FPL Group. Other independent developers are companies that were originally started by utility holding companies, and have been, or are in the process of being, spun off into successful stand-alone companies. These include Mirant (formerly part of Southern), Reliant Resources (Reliant Energy), and NRG (Xcel Energy).

Last, a group of players has been traditionally associated with various aspects of the oil and gas industry that have now diversified into power generation. These include companies such as Enron, Dynergy, Williams Energy, El Paso, and Kinder Morgan.

An important, but sometimes overlooked, fact about independent power plant developers is that they, and their shareholders, incur the risks associated with their power plant investments. The rewards and penalties possible for incurring these risks are a double-edged sword. Investments in tight generation markets that yield high returns are clearly a benefit that is

Table 6.1 Top 25 U.S. Power Plant Developers

Rank	Company	Minimum (MW)	Maximum (MW)	Minimum Percent of Total	Maximum Percent of Total
1	Calpine Corp.	30,186	31,283	15.9	15.9
2	Duke Energy	17,537	17,755	9.3	9.0
3	Cogentrix	12,265	13,431	6.5	6.8
4	Panda Energy	12,236	12,406	6.5	6.3
5	PG&E Corp.	12,202	12,202	6.4	6.2
6	Mirant Corp.	8866	9519	4.7	4.8
7	PSE&G	8760	8810	4.6	4.5
8	FPL Group	8441	8645	4.5	4.4
9	International Power	8291	8881	4.4	4.5
10	Tenaska	8146	8246	4.3	4.2
11	Constellation Energy	6582	7136	3.5	3.6
12	Southern Company	6084	6094	3.2	3.1
13	AES Corp.	5780	6285	3.1	3.2
14	Reliant Energy/Resources	5621	5678	3.0	2.9
15	TECO Energy	5473	5758	2.9	2.9
16	Xcel Energy/NRG	4923	4930	2.6	2.5
17	Enron Corp.	4025	4134	2.1	2.1
18	PPL Corp.	3938	4060	2.1	2.1
19	Dynergy Inc.	3928	4058	2.1	2.1
20	Progress Energy	3465	3519	1.8	1.8
21	El Paso Corp.	3285	3290	1.7	1.7
22	Kinder Morgan	3019	3019	1.6	1.5
23	Allegheny Energy	2338	2338	1.2	1.2
24	Exelon Corp.	2012	2189	1.1	1.1
25	Orion	2000	2738	1.1	1.4
	Total	189,403	196,404		

Source: From Ellinghaus, C., *U.S. Electricity Supply & Demand Analysis: Tight Gas Supply Tells the Story,* New York, Williams Equity Research, 2001. With permission.

misunderstood as an exercise of market power. One needs to only look at the reactions to the current California crisis as an indicator of how surrealistic the misperceptions of these market risks can be perceived.

What is often not considered is the probability that independent providers could also incur losses associated with their investments when markets become saturated with large numbers of highly efficient and low-cost power plants. In cases like these, independent providers and their shareholders will bear 100% of the risks associated with these failed investments. Such risks, and the participants who bear them, are in stark contrast to the stranded cost problem for traditional monopoly utilities during the retail choice process. In most instances, ratepayers were required to pay all, or most, of the costs of these uneconomic investments.

6.4 Analyzing the investment opportunity

It would be instructive to analyze the potential investment opportunity presented in operating a merchant plant at this particular time in the deregulated wholesale market for electricity. Assume we have the opportunity to site a 155-megawatt (MW) plant with an estimated construction cost of $50 million in the Midwest. Our business strategy is to cover all fixed costs using a long-term contract with an anchor tenant and sell any excess capacity on the spot market where the average price is $30/MWh. To determine an appropriate anchor tenant whose demand would allow us to break even, we perform a basic cost–volume–profit analysis. This analysis uses sales price, fixed cost, and variable cost per unit to determine pretax profitability at various levels of output (in this case, MWh). Table 6.2 presents the calculation of fixed costs for a hypothetical independent power provider (IPP) facility.

The prime variable cost for operating the facility is the cost of natural gas used to move the simple cycle turbine. Variable cost is the cost of natural gas delivered to the plant. This cost is estimated as $0.033/kilowatt-hour (kWh) for a normal gas turbine, less a 40% efficiency gained with the combined-cycle plant. The variable cost is thus estimated to be $19.80/MWh.

To develop a bid for the anchor tenant, assume the average price for industrial customers is $46.70/MWh. Assume that because the tenant is close to the facility there are no transmission charges. To ensure competitiveness,

Table 6.2 Assumed Fixed Cost for Hypothetical IPP

Fixed Cost Amount	Assumption
$1,670,000	Straight-line depreciation at 30-year life
$1,300,000	Labor costs
$1,600,000	Total debt service; 40% debt at 8%
$4,570,000	Total fixed costs

the bid is submitted as low as $30/MWh, a price that approximates the wholesale price on the spot market. The price bid will also be tied to a long-term contract and adjusted with the market price of natural gas.

The break-even point can be calculated as follows:

Break-even (MWh) = Fixed Costs/(Selling Price – (Variable Cost/Unit)
= $4,570,000/($30.00 – $19.80)
= 448,040 MWh

Assuming a reliability of 95%, 1,289,910 MWh are available during the year (24 × 365 × 0.95 × 155).

Therefore, the break-even utilization is 448,040 MWh/1,289,910 MWh, or 34.7%, which is 53.8 MW of capacity that needs to be tied down on long-term contract. We look for a target anchor tenant that has a demand for 60 MW. With the anchor tenant under contract, we have covered our fixed cost of operation and have 95 MW to sell on the spot market. The following is an estimate of what we could have earned with this capacity in the Midwest during the summer of 1998. The May–August average-weighted (by MWh) price into the Cinergy hub was $78/MWh.[4] If execess capacity had been sold competitively at this price, how much money would have been made?

Pretax Profit = (Selling Price – (Variable Cost/Unit)) × MWh available
= ($78.00 – $ 19.80) × (123 days × 24 hr/day × 0.95 × 95 MW)
= $15,505,528, or an annual return of 31% on the $50 M investment for only months of operations

6.5 *Transmission issues associated with IPP development*

The electric transmission grid is an important means by which power is moved between regions. The grid not only facilitates physical power flows, but it assures that competitive transactions between regions are possible. As a result, the grid is very important in promoting competition. Plants that cannot secure available transmission capacity to move their power will be limited in their market opportunities.

The power transmission grid facilitates competition in two important manners. When regional wholesale price differentials exist, transmission can serve as the means of equalizing these differentials as cheaper power moves to more expensive regions until prices between the two areas are close to equal. This movement assures that the "law of one price" will be closely approximated.

The second important role that the transmission system can play is in minimizing market power in a particular region. Consider, for instance, an incumbent utility that because of its past role as monopoly provider of utility services owns a significant amount of regional generating capacity. It would be difficult for that incumbent utility to exercise market power if power from other resources, in other regions, were able to flow into the region and undercut the potential market power pricing abuses of the incumbent.

The problem with the transmission system in the current competitive wholesale market, however, is twofold. First, the electric power transmission system has been developed over a number of decades under traditional utility regulatory practices and policies. In the past, the interrelated system of individual transmission systems was developed for reliability purposes. For instance, if one region found itself short on electrical generating capacity, it could draw upon the resources of a neighboring utility to meet that shortfall.

Over time, economic considerations entered into the picture as economy energy sales between utilities began to take place. These sales would be made when incremental generation by one neighboring utility was less than that of another. Consider two hypothetical utilities called Utility A and Utility B. If Utility A had generating capabilities that were more cost effective, at the margin, than those of Utility B, these two utilities would have opportunities for trade. In the past, these trades were limited and were usually made on a "split-the-savings" basis. For instance, if Utility A had marginal costs of $25/MWh and Utility B had marginal costs of $30/MWh, then Utility B would ramp down its generation and purchase the cheaper resources. The differential ($5/MWh) would be shared between the two utilities (i.e., $2.50/MWh apiece).

However, in the past, these opportunities for trade were somewhat limited, and the traditional way to meet demand over the long term was to build new generating facilities. Thus, while some trade has existed over the past several decades, it was very limited in nature and did not place commercial and physical strain on the use of the power transmission system.

This paradigm, however, shifted with the advent of Order 888 and wholesale competition. With Order 888, new and higher volumes of trade began to move between and across regions. This placed pressure on the physical operation, pricing, planning, and organization of the utility transmission network. One means of addressing these pressures was to organize the utility transmission system under an organization referred to as an Independent System Operator, or ISO. The advent of wholesale competition saw a strong preference for the idea of an ISO. However, questions about the operating incentives of ISOs have given rise to debates over independent transmission companies or, transcos, and alternative methods for transmission system governance.

ISOs are one of the earliest proposed forms of transmission governance to be facilitated in restructured markets. FERC, in Order 888, gave a strong preference to the ISO concept and its principles. ISOs are essentially nonprofit organizations that work like independent air traffic controllers for a given regional transmission system. While ownership of transmission systems stays with utilities, ISOs take over the security and operational control over all power flows and wholesale transactions. These entities, for the most

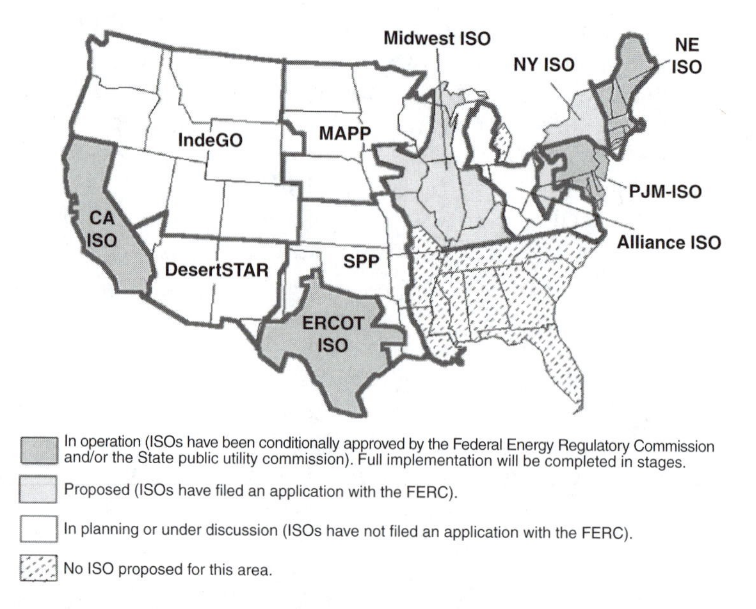

Figure 6.3 Independent system operators in operation, proposed or under development (March 1998). (From U.S. Department of Energy, Energy Information Administration, Electric Power Annual, 1999.)

part, were either in control or directed long-term planning and pricing regimes for the regional utility transmission members.

In order to assure true independence, neither ISOs nor their employees are allowed to have any financial interest in the transmission system, its operation, or the transactions occurring over the system. An ISO has an independent governing board that includes not only utility representatives, but also representatives from other stakeholder groups including power marketers, independent power producers, small customers groups and, in some instances, environmental groups. The open, objective manner of transmission operation has led ISOs to be a preferred method of transmission organization, as seen in Figure 6.3.

ISOs have been plagued by their detractors from the onset of the electric restructuring debate. One initial criticism laid upon the formation of ISOs rested with the enormous costs associated with creating a new bureaucracy to manage regional transmission grids. The experiences and costs associated with the creation of the California ISO and its associated power exchange (PX) provided justification for this criticism. Others argued that ISOs did not go far enough in removing incentives for cross-dealings and potentially preferential treatment. However, one of the most significant criticisms leveled against the ISO ideal rests with concerns about its short- and long-run incentives as a nonprofit organization.

ISO critics have questioned the motivations of nonprofit organizations to plan for and manage the transmission system efficiently. This system will continue to be owned by utilities that have a fiduciary responsibility to their shareholders to maximize the profits that could be earned on these assets. However, a nonprofit organization will be removed from fiduciary responsibility, and may even act at cross-purposes with utility motivations for maximizing shareholder returns.

For instance, ISOs, it is argued, will have little or no incentive to reduce costs, introduce new technologies, or make management and operating innovations. The inability to earn profits could make ISOs relatively indifferent to such long-run planning issues as increasing transmission capacity or making substation upgrades and additions. The lack of incentives has led many critics, primarily transmission-owning utilities, to call for an alternative means to organize and govern the transmission system.

One of the more recent proposals for transmission organization rests with transcos. The transco idea attempts to merge the concepts of independence and inclusiveness of an ISO with the profit-maximizing goals of a private enterprise. Recent transco proposals envision a private corporation that would operate and manage utility transmission assets on a for-profit basis. The owners of these assets, in turn, would serve as shareholders in this new corporation. Management of a transco would then be accountable to their shareholders. Transcos would be for-profit entities, but could include membership and nonvoting input from nontransmission-owning stakeholders such as municipal utilities, rural distribution cooperatives, power marketers, and independent power producers.

While transcos have appeared to become the preferred approach for encouraging investment in the transmission system, securing independent governance across regions, particularly the Gulf South, has been a more challenging issue. Figure 6.3 shows that, even after Order 888, the southern part of the United States avoided the trends in regional transmission governance and became balkanized into a system of unorganized entities run, or partially controlled, by incumbent transmission-owning utilities.

The challenge for federal regulators has been to encourage development of independent organizations and to do so in a manner broad enough in scope to secure independence, as well as potential operating efficiencies across regions. In a recent order, FERC took its boldest stand on the issue by forcing all parties to the table for 45 days of negotiations to bring the U.S. power transmission system into five major systems: West, South, Northeast, Midwest, and Texas. These systems will be organized into large regional transmission organizations (RTOs) that will handle a variety of different transmission operation, pricing, and planning issues. While it is still too early to tell, the promise of having a number of large regional RTOs, with a number of for-profit transcos, seems likely.

Another issue associated with the nexus between merchant power and transmission is how these competitive generators of electricity facilitate the power system. A common misperception about merchant generation is that it somehow gets a free ride on the transmission system. The argument that independent power plants somehow exploit the existing transmission system ignores a number of important technical and regulatory considerations.

First, when new generation or new load is added to a transmission system, the flows on the system change. The proper siting of new generation on the system can often eliminate the need for transmission upgrades and maximize the capability of the transmission system as a whole. For example, one location on the transmission system may be experiencing line overload or congestion, while another location may be experiencing low voltage. This problem could be solved by either building additional transmission to strengthen the grid or by strategically locating additional generation on the system. This additional generation would change load flow on the transmission system, improve voltage profiles on the system, and enhance overall reliability.

Second, FERC's current policy for assigning costs for transmission services is summarized in its "Inquiry Concerning the Commission's Pricing Policy for Transmission Services Provided by Public Utilities under the Federal Power Act; Policy Statement." This policy requires, among other things, that rates for transmission services must ensure that "costs incurred in providing the wholesale transmission services ... are recovered from the applicant ... and not from existing wholesale, retail, and transmission service customers." This policy is contained in the current pricing rules for new generator interconnections and new requests for transmission service. Therefore, existing retail customers can be assured they will not be negatively impacted from a rate standpoint by the entry of new generation on the transmission grid.

Third, independent power providers exist to take advantage of unique cost and demand characteristics in particular regions. The profit motive serves end users well, because as more of these generators enter a particular region, they displace older, less-efficient generating units and/or supplement the regions' existing generating resources. However, to maximize the profit opportunities for these facilities, trade between regions must be facilitated. A number of antipower plant activities have proposed to restrict sales of merchant providers to a particular region. This can change the profit dynamics of the facilities and could discourage certain generating projects. Merchant plants are no different than other large industrial and manufacturing facilities. If an automobile manufacturer were to locate in Louisiana, not all, or some significant portion, of its output would be required to be sold in the state. It seems unreasonable to expect the same from an independent power facility.

6.6 Conclusions

Wholesale markets have responded vigorously to the competitive signals sent by federal energy regulators in Order 888. The plethora of new generating facilities across the United States is clear evidence of this response. Despite this response, these wholesale markets have, and will continue, to experience a number of transition problems. These problems are almost exclusively restricted to transmission planning, operation, and governance. Comprehensive and stable policies are yet to be developed, despite the best attempts of federal regulators. To date, the inertia of past principles and control philosophies has been too great.

Bulk power markets are traditionally defined as including both transmission and generation assets. In the past, both resources were utilized to meet end-user power requirements. While the existing regulatory paradigm has separated these two sectors of the industry, their interdependence is still being felt. To ensure efficient market outcomes, reconciling the conflicts between greater power generation and trading and stability and long-term transmission planning will need to addressed.

References

1. Daly, M., Federal Reserve Bank of San Francisco Economic Letter, Regional Report, April 20, 2001.
2. Energy Information Administration, The Changing Structure of the Electric Power Industry, Washington, D.C., U.S. Department of Energy, 1993: 21.
3. Staff Report to the Federal Energy Regulatory Commission on Western Markets and the Causes of the Summer 2000 Price Abnormalities, Part 1 of Staff Report on U.S. Bulk Power Markets, Washington, D.C., Federal Energy Regulatory Commission, 2–3.
4. Staff Report to the Federal Energy Regulatory Commission on the Causes of Wholesale Electric Price Abnormalities in the Midwest During June 1998, September 22, 1998, Figure 3-3, Washington, D.C., Federal Energy Regulatory Commission, 3–7.

chapter seven

Understanding both technical and business factors

7.1 A brief history

Toward the end of his career, Thomas Edison was asked, "What was your greatest invention?" In response, he said, "incandescent electric lighting and the power system." Edison's answer indicates that he saw the importance of technological innovation, but was also an entrepreneur. He understood that a complete electrical power system would be required to make incandescent electric lighting useful and enable competition with gas lighting companies. The idea of competition was part and parcel of the electric power industry from its very inception.

Thomas Edison's Pearl Street power system in New York City became fully operational in 1882. The Pearl Street power system is sometimes cited as the first electric power system. This power system was different from modern power systems in several respects:

1. The Pearl Street system was a direct current (D.C.) system.
2. All of the power generation facilities were in a single location.
3. The low-voltage power delivery system was entirely underground.
4. The system provided electricity for a single application: street lighting.

After a debate between Thomas Edison and George Westinghouse about the relative merits of alternating current (A.C.) systems vs. D.C. systems, the architects of the early power systems finally decided to use three-phase A.C. generators with step-up transformers, high-voltage transmission systems, and step-down transformers at the point of customer service. Three-phase A.C. systems delivered a constant power supply and reduced transmission losses, but required synchronizing generation units as they were added to the system. The need to cope with increased system operating complexity to reduce power supply costs has always been recognized by power system engineers.

During the early part of the 20th century, more and more generators were added as interconnected systems expanded, and it became necessary to develop methods for coordinated system operations. During this era, the primary tools for power system operations were a frequency meter and a telephone. The frequency meter was used to determine if the amount of aggregated power generation output was greater than, less than, or equal to the instantaneous total load. Telephones were used to communicate with plant operators so that unit outputs could be increased or decreased to balance total system load with total system generation.

Modern power systems are A.C. systems consisting of many power generation facilities operating in parallel connected by a vast high-voltage transmission network. And, of course, modern power systems provide a platform for many uses, including lighting, heating, air conditioning, computers, VCRs and other electronic devices, the electrical motors used in industry, and many other applications.

7.2 The current situation

Today, more than ever before, those who design and operate electric power systems must understand both the technical and business factors involved. Today it is necessary to understand:

1. Industry standards relating to electrical safety
2. North American Electric Reliability Council (NERC) reliability practices and standards
3. The laws and regulations relating to competition and open access
4. The business environment and NERC business practice standards
5. The laws and regulations relating to environmental protection
6. Community concerns relating to environmental protection

7.3 Industry standards relating to electrical safety

The United States, unlike other countries, has traditionally depended on voluntary standards processes. Traditionally, the U.S. standards-making system has been coordinated by The American National Standards Institute (ANSI). Public and private interests in standards have been well served by the ANSI process. ANSI has assured that the standards-developing process has been open, balanced, based on consensus, and has followed a "due-process approach" (providing opportunities for review, comment, and redrafting).

The two standards documents of foremost importance from an electrical safety point of view are the National Electrical Safety Code (NESC) and the National Electrical Code (NEC). The National Electrical Safety Code focuses on electric utility practices. The following is from the NESC:

> The purpose of the NESC is the practical safeguarding of persons during the installation, operation, or maintenance of electric supply and communication lines and associated equipment. The NESC contains the basic provisions that are considered necessary for the safety of employees and the public under the specified conditions. The NESC covers supply and communication lines, equipment, and associated work practices employed by a public or private electric supply, communications, railway, or similar utility in the exercise of its function as a utility. It covers similar systems under the control of qualified persons, such as those associated with an industrial complex or utility interactive system.

The National Electrical Code (ANSI/NFPA-70) focuses on customer installations. The following is from the NEC:*

> The National Electrical Code, NFPA-70, addresses proper electrical systems and equipment installation to protect people and property from hazards arising from the use of electricity in buildings and structures. The NEC covers: 1) Installations of electric conductors and equipment within or on public and private buildings or other structures, including mobile homes, recreational vehicles, and floating buildings; and other premises such as yards, carnivals, parking lots, and industrial substations. 2) Installations of conductors and equipment that connect to the supply of electricity. 3) Installations of other outside conductors and equipment on the premises. 4) Installations of optical fiber cable. 5) Installations in buildings used by the electric utility, such as office buildings, warehouses, garages, machine shops, and recreational buildings that are not an integral part of a generating plant, substation, or control center.

The NEC and the NESC were written as voluntary standards. Some portions of these codes have been subsequently adopted by some local authorities and nonadherence, in some cases, is a violation of state laws. Furthermore, the National Society of Professional Engineers (NSPE) Code of Ethics points out that professional engineers are expected to conform to applicable standards and "hold paramount the safety, health, and welfare of the public."

Electrical equipment suppliers are increasingly functioning in global markets and, consequently, international standards activities are of increased

* National Electrical Code

importance. Currently, ANSI is the official U.S. representative to the International Organization for Standardization (ISO) and, via the U.S. National Committee, the International Electrotechnical Commission (IEC). The IEC and the ISO are the primary international standards-making organizations of interest to the electric power industry.

Some believe that U.S. involvement in the development of international electric power systems standards may be reduced as one of the consequences of electric industry deregulation. In an effort to be more competitive, some companies have cut back on their budgets for travel to standards meetings, particularly to international standards meetings. Corporate downsizing, also caused by efforts to become more competitive, has also affected participation in standards-making activities.

7.4 NERC reliability practices and standards

NERC was formed in 1968. The need for increased reliability coordination was recognized and justified as a consequence of the Northeast Blackout of November 9, 1965.

The Northeast Blackout caused more than 30 million people to be without power for a long period of time, in some cases as long as 13 hours. It was, and still is, the most extensive electrical power blackout that North America has ever experienced.

President Lyndon Johnson described the Northeast Blackout in a letter written to the chairman of the Federal Power Commission. The following excerpt and more details concerning the 1965 Blackout are available on the Central Maine Power Company's website (http://www.cmpco.com).

> Today's failure is a dramatic reminder of the importance of the uninterrupted flow of power to the health, safety, and well being of our citizens and the defense of our country.
>
> This failure should be immediately and carefully investigated in order to prevent a recurrence.
>
> You are therefore directed to launch a thorough study of the cause of this failure. I am putting at your disposal full resources of the federal government and directing the Federal Bureau of Investigation, the Department of Defense, and other agencies to support you in any way possible. You are to call upon the top experts in our nation in conducting the investigation.
>
> A report is expected at the earliest possible moment as to the causes of the failure and the steps you recommend to be taken to prevent a recurrence.

The Northeast Blackout was initiated when relays opened a transmission line near the Niagara Falls generating facility. Opening this line reduced power flows from the generating facility and caused several other lines to be overloaded. After this, "cascading outages" occurred in Ontario, New York State, New Jersey, Pennsylvania, and parts of New England.

After the Northeast Blackout, U.S. electric utility industry executives recognized the need for a "National Electric Reliability Council" (NERC). NERC was initially established as a nonprofit, voluntary organization owned by Regional Reliability Councils. At the outset, utility managers and utility technical experts were invited to participate to serve the mutual self-interests of those involved. The NERC evolved into the North American Electric Reliability Council in recognition of the fact that the electric power system of the United States was integrated with the electric system in Canada and part of the Mexican power system. NERC hired a small staff located in Princeton, NJ, and structured its committee system to include an engineering/planning committee and an operating committee.

The efforts of NERC and its members have helped to make the North American electric system the most reliable electric system in the world. NERC has served as a forum for information exchange, developed operating and planning standards for its members to follow, reviewed planned generation and transmission systems, studied past electric system disturbances, and provided education and coordination for various groups. NERC played an important coordinating role with respect to "Y2K" concerns.

NERC is now in the process of dramatically changing its structure and operations to address the profound changes taking place in the structure and operations of the electric power industry. NERC has restructured its board of trustees to include all segments of the electric industry, including investor-owned utilities; federal power agencies; rural electric cooperatives; state, municipal, and provincial utilities; independent power producers; and power marketers. The NERC Regional Councils, now ten in number, have also opened up their membership to include all of the industry stakeholders and independent participants.

In 1996, NERC began formalizing its transmission operations policies and called for the establishment of regional security coordinators to proactively monitor system conditions and mitigate potential reliability problems. Once security coordinators were identified by the region, NERC formed a security coordinator working group. The 22 members of the security coordinator working group developed improved procedures for interregional coordination.

NERC has charged security coordinators with seeing the big picture, assessing the moment-to-moment reliability of the grid, taking actions necessary to maintain reliability in the best interests of the interconnection, and being responsible for coordination during emergencies. Security coordinators have a central role in maintaining reliability.

NERC has recognized that a voluntary system of compliance is no longer adequate and has begun transforming itself into a new organization to be called the North American Electric Reliability Organization (NAERO). NERC has established a new compliance process. Under the new process, compliance with NERC procedures is mandatory, not voluntary. However, in the absence of federal legislation and statutory authority, NERC is currently unable to enforce compliance. The industry is working toward the passage of legislation to correct this situation.

7.5 Laws and regulations relating to competition and open access

The laws and regulations relating to competition and open access have been discussed at some length in the other chapters of this book.

7.6 The business environment and NERC business practice standards

In order to implement the laws and regulations relating to competition and open access, it has been necessary for the electric power industry to expand the role of the North American Electric Reliability Council. In 1998 NERC formed a "Market Interface Committee" (MIC) to review NERC planning and operating policies. NERC charged the MIC with determining any impacts the planning and operating policies might have on commercial practices and the impacts market practices may have on reliability.

7.7 End-of-chapter questions

1. Was the concept of "competition" given any consideration when Thomas Edison designed and operated the Pearl Street power system in 1882? Explain.
2. What were some of the major technical differences between the 1882 Pearl Street power system and modern North American power systems?
3. In the current business environment, is it necessary for an engineer to keep up with new industry standards and public policy developments, or is it sufficient to have technical expertise in a single area of specialization? Give an example of how an engineer could have been "overtaken by events" if he or she had only focused on the technical aspects of his or her job.
4. What is the purpose of the National Electrical Code?
5. What is the purpose of the National Electrical Safety Code?
6. What event prompted the electric power industry to form the North American Electric Reliability Council (NERC)?
7. How is the role of the North American Electric Reliability Council changing in response to changes in the industry?

chapter eight

The North American bulk electric system

8.1 The evolution of system operations and control

Understanding how the methods for power system operations and control have evolved during the last century can provide insight into the methods used for the operation of modern power systems. The pioneering efforts of power system operators (or dispatchers) and power engineering innovations have made it possible for North American electric systems to achieve a very high level of efficiency. Today the bulk electric systems in North American are the most reliable systems in the world. Whether this level of reliability will continue during and after the transition to "restructuring and deregulation" is an interesting subject for discussion.

Central dispatching systems were not used in the first power systems from the early 1880s to the early 1920s. Generation control was accomplished at power plants by local equipment. The Philadelphia Electric Company in Pennsylvania installed one of the first central dispatch generation control systems in 1923. At that time, power systems were still operated as "islands," that is, there were no interconnecting tie lines and, consequently, no wholesale power sales between electric utility systems.

The first interconnections between electric utility systems were not constructed until later in the 1920s, also in the Pennsylvania-New Jersey-Maryland area (now referred to as the PJM system). Having physical tie lines or interconnections between power systems provided advantages to system operators, but also introduced new operating complexities. The advantages of interconnections are that they permit sharing generation reserves during emergency conditions and allow interconnected electric systems to make economic transactions when load diversities and generation scheduling plans create opportunities. To obtain these benefits, electric systems are required to coordinate their operations. Coordination initially involved basic control concepts, such as the "load-balancing function" and "time-error correction." However, coordination has become increasingly more complex and now requires very elaborate procedures for monitoring "inadvertent interchange," "frequency response characteristics," and "area interchange errors." Of course, as coordination has become more complex, the operating agreements and operating manuals have become more voluminous and much more sophisticated.

Ineffective coordination, or a lack of compliance with operating agreements, has always had an adverse effect on operating economics and/or power system reliability. During most of the 75-year history of interconnected operations, most people, including most electric utility company employees, have paid very little attention to these impacts. Prior to the advent of deregulation and restructuring, only system operators had an intimate knowledge of the actual impacts of a lack of coordination. Today, there are many participants in electric power markets, and many of these participants are closely monitoring the operating actions taken and whether these actions are consistent with operating agreements.

New industry organizations were formed to coordinate power system operations as the number of physical interconnections increased. One of the first such organizations was the Interconnected Systems Group formed in 1933. Many renamings and recombinations have occurred during the last seven decades, and more are yet to come. Reliability councils and operating committees have been formed based on geographic proximity and similar operating philosophies. Today the following ten regional reliability councils fall under the North American Electric Reliability Council (NERC) umbrella:

1. ECAR: The East Central Area Reliability organization is located in nine East-Central states of the United States.
2. ERCOT: The Electric Reliability Council of Texas is located entirely within the state of Texas and serves most of the electrical demand of the state.
3. FRCC: The Florida Reliability Coordinating Council is located entirely in the state of Florida and serves most of the electrical demand of the state.
4. MAAC: The Mid-Atlantic Area Council is geographically the same as the PJM control area and serves electrical demands in Virginia, Pennsylvania, New Jersey, Maryland, Delaware, and Washington, D.C.
5. MAIN: The Mid-America Interconnected Network serves electrical demands in all of Illinois and portions of Missouri, Wisconsin, Iowa, Minnesota, and Michigan.
6. MAPP: The Mid-Continent Area Power Pool serves electrical demands in Minnesota, Nebraska, North Dakota, Manitoba, Saskatchewan, and parts of Wisconsin, Montana, Iowa, South Dakota, Kansas, and Missouri.
7. NPCC: The Northeast Power Coordinating Council serves electrical demands in New York, the six New England states, Ontario, Quebec, New Brunswick, Nova Scotia, and Prince Edward Island.

8. SERC: The Southeastern Electric Reliability Council serves electrical demands in parts of Virginia, the Carolinas, Kentucky, Tennessee, Texas, Louisiana, Mississippi, Arkansas, Georgia, and Florida.
9. SPP: The Southwest Power Pool serves electrical demands in parts of Oklahoma, Missouri, Kansas, Colorado, Texas, Louisiana, Arkansas, and Mississippi.
10. WSCC: The Western Systems Coordinating Council serves electrical demands in parts of Arizona, California, Colorado, Idaho, Montana, Nebraska, Nevada, New Mexico, Oregon, South Dakota, Texas, Utah, Washington, Wyoming, Alberta, British Columbia, and the northern portion of Baja California Norte, Mexico.

8.2 *The big machines*

From an electrical point of view, there are three systems in North America. These systems are called "interconnections." They are:

1. The Western Interconnection, generally including the states west of the Rocky Mountains and the Western Canadian provinces.
2. The ERCOT Interconnection, including most of Texas. ERCOT does not have synchronous interconnections to other states and is not under the jurisdiction of the Federal Energy Regulatory Commission (FERC).
3. The Eastern Interconnection, including the Eastern Canadian provinces and most of the United States, east of the Rocky Mountains.

Each one of the Interconnections is comprised of a group of loads, transmission systems, and generators operating in synchronism. The Interconnections are not connected to each other by synchronous interconnections. However, there are D.C. tie lines between the Eastern Interconnection and ERCOT, and between the Eastern Interconnection and the Western Interconnection.

Interconnections can be viewed as a single machine consisting of many synchronous elements. An interesting debate has recently been initiated on the engineering community's "power globe" comparing the two "very large machines" that have been created during the last century: the electric power system and the World Wide Web. The question is, "Which is the most complex machine?" It has been noted that from a purely financial point of view, the investment in equipment and the annual revenues are greater for the electric power system. The presumption is that the more expensive machine is more complex.

On the other side of the coin, current average starting salaries for computer system/Internet engineers are higher than current average salaries for power systems engineers. Here the presumption is that the more complex machine requires higher-paid engineers. As a practical matter, we should probably not be concerned about separating the two big machines, because deregulation and restructuring are causing the two machines to merge into a single machine. The computer technology and the Internet are increasingly used for the management and control of the power system, both to maintain system reliability and to facilitate real-time markets.

8.3 End-of-chapter questions

1. One disadvantage of having an extremely reliable bulk electric system is that operators gain little experience with restoration techniques. Also, in some cases, U.S. electric utility executives have been criticized for spending too much money and "gold plating" their systems. Do you think that many electricity customers would be willing to accept a lower level of reliability for slightly reduced rates?
2. Is the regional reliability council of NERC organized so that each state belongs to a single regional council, or is the organization based on some other system? Explain.
3. How is an Interconnection defined? Note that there is a difference between an "interconnection" (lower case) and an "Interconnection" (upper case).
4. How many "Interconnections" are there in the United States? Geographically, which is the largest of the Interconnections?
5. Why are the power systems in ERCOT the only power systems in the United States that are not under FERC's jurisdiction?
6. Is it possible to move power from one Interconnection to another Interconnection? If so, how is this accomplished?

chapter nine

Methods for economically operating a power system

9.1 Operating economics, control systems, and power systems reliability

We have devoted one chapter in this book to the subject of operating economics, as indicated by the title of this chapter. This was done to focus on key concepts and to facilitate teaching using this book. However, in the "real world" (or the physical world), it is impossible to separate operating economic considerations from power system control considerations. Additionally, power system operating or control decisions involve taking into consideration possible impacts on power system reliability and environmental performance.

Possibly the best way to begin to understand operating economics and the related issues of control, reliability, and environmental performance is to begin by considering a power system consisting of a single generating unit attached to a single variable load. After this, a power system with multiple generating units is considered, and the classical "economic dispatching" problem and the classical "unit commitment" problem defined. These problems are presented both with and without consideration of transmission losses. Finally, to make this treatment more realistic, the additional complexities of allowing power purchases and sales between interconnected power systems are considered. This level of understanding is necessary to understand the operating economics/systems control approaches used by the industry prior to deregulation and restructuring. Other chapters of this book extend this discussion to the new business environment in which open access and power marketing considerations must also be taken into account.

9.2 A single generating unit

First, a power system consisting of a single generating unit attached to a single variable load (or a set of loads which aggregated constitute a single variable load) is considered. Most of the generating capacity used in the

United States involves burning coal to produce steam and turn the shaft of an alternating-current, synchronous generator. Since this is the most common technology, assume that the single generating unit coal-fired. Of course, from an electrical point of view, the fuel source (or heating source) is not important. Any steam power plant (coal-fired, gas-fired, nuclear, etc.) produces steam that impacts the turbine blades causing the shaft in the synchronous generator to rotate. And, of course, we know from physics that voltage is induced in the coils of the generator in the presence of the generator's magnetic field.

For the purposes of this discussion, assume that the load is connected directly at the terminals of the generator. In other words, there are no transformers, transmission or distribution lines, switch gear, protective devices, etc. Of course, in actual power systems, power generators generally have three-phase output voltages in the range of 10 to 24 kV, and it is necessary to use step-up transformers to reduce the losses associated with long-distance transmission. Also, loads are generally at much lower voltages than transmission-level voltages, and, therefore, step-down transformers are required. The use of system protection equipment and other substation or electrical system equipment is outside the scope of this discussion, except to say that its proper operation plays an important role in keeping the power system reliable, thereby enabling economic operations.

The single generating unit problem is almost trivial. If there is no control over the load, the only option is to match the output of the generating unit to the load level. As long as the load level is greater than the unit's minimum possible output and less than the unit's maximum possible output, it is possible to serve the load.

If there is control over the amount of load, the problem becomes slightly more complex. In this case, it is possible to find an optimal level of generating output from an economic point of view. The methods for doing this are in most power systems analysis textbooks. It is normally assumed that the relationship between fuel input (f) and power output (P) can be expressed with an equation of the form:

$$f = a\,P^2 + b\,P + c.$$

Of course, using fundamental calculus, the minimization of this function is accomplished by taking the first derivative and setting it equal to zero. The first derivative of this function is usually called the "incremental cost."

9.3 Two generating units

When two generating units are connected together to serve a single variable load (or a set of loads that aggregated constitute a single variable load), a decision has to be made. Should one unit's output be held to a single value and "follow load excursions" with the other unit, or should the power output levels of both units be varied as load varies?

In the early days of interconnected operations, it was common practice to use a single unit for load following and to accomplish frequency regulation. As system operations techniques became more sophisticated, it was recognized that this would not result in an "optimum economic dispatch." Again, recognizing that the relationship between fuel input and power output for each unit (where the units are numbered from i = 1 to i = N) can be expressed with an equation of the form:

$$f_i = a\,P_i^2 + b\,P_i + c.$$

Again, taking the first derivative for each unit, the incremental costs for each unit can be found. Power systems analysis textbooks show that, ignoring transmission losses, optimal economic dispatch is achieved when the incremental costs for all units are equal.

This principle has been implemented in modern energy-management centers and has resulted in billions of dollars of production cost savings.

9.4 End-of-chapter questions

1. In what ways are coal-fired power plants different from nuclear power plants? In what ways are they similar?
2. If four identical generating units are in a single power system, and each has a minimum output of 10 MW and a maximum output of 200 MW, how much system load can be served? (Take service reliability into account, but do not consider economic factors at this point.)
3. Using the same four units in question 3, how would you approach the problem of finding the amount of load that could be served if it is desired to minimize fuel costs?
4. Ignoring transmission losses, what is the condition for optimal economic dispatch from a group of units that do not necessarily have the same input/output characteristics?

chapter ten

Power generation control

10.1 The definition of automatic generation control

Automatic Generation Control (AGC) is a means of automatically controlling the outputs of power-generating units to accomplish economic dispatch, and maintain system frequency and power flows over tie lines at desired levels. AGC, sometimes referred to as load control or load frequency control, is performed at energy control centers or energy coordination centers using energy management systems. Energy management systems acquire data from the power system and use computers to process the data. Modern energy management systems usually have sophisticated provisions for operator interaction and include the equipment and communications required to send control signals to generating units.

AGC supplements the local control that occurs at power plants. At thermal generating plants, local control systems regulate turbine-generator speed by responding to changes in system frequency and adjusting steam flows to increase unit power outputs when system frequency is low or decrease unit power outputs when system frequency is high. Speed regulation is also called "governor droop." Governor droop is defined as the percent change in frequency that would cause the unit's generation to change by 100% of its capability.

Traditionally, AGC has been implemented as a simple feedback control system in which the error to be driven to zero is defined as having two components. The first component recognizes differences that may exist between actual and scheduled tie flows, and the second component accounts for deviations from scheduled frequency. The AGC error signal, called area control error (ACE), is mathematically defined as:

$$ACE = (P_a + P_s) - 10\ B_f\,(f_a - f_s),$$

where P_a is the actual net interchange power over the system's tie lines; P_s is the scheduled net interchange power over the system's tie lines; B_f is a frequency bias constant; f_a is the actual system frequency; and f_s is the scheduled system frequency.

ACE, P_a, and P_s are usually expressed in MW; f_a and f_s are expressed in Hz; and B_f is normally expressed in MW/0.1 Hz.

AGC is currently accomplished in the United States in 136 control areas. It should be noted that control areas have responsibility not only for AGC, but also for scheduling power transactions. The scheduling function is becoming increasingly important in the new era of industry deregulation and restructuring. Combining control areas creates greater opportunities for optimizing control and for scheduling economic power transactions. During the last several years, there have been many corporate mergers among electric utility companies. Corporate mergers often result in more efficient manpower utilization, but may also result in combining control areas. It will be interesting to see how many control areas will result from the restructuring activities undertaken in this decade.

A more complete description of the role of AGC as it relates to economic operations may be found in most of the power system analysis textbooks published during the last 2 decades.

There can be significant cost implications from inappropriately setting the frequency bias constant, B_f, especially during the "peak" load period. Carefully tuning the AGC system is well justified by the very large economic savings that can be realized through interconnected operations, as compared to control areas operating as isolated entities.

AGC regulates system frequency under normal conditions, but it does not play a role in limiting the degree of frequency deviations that occur within seconds after a major system disturbance, such as the loss of a major generating unit or the tripping of a large block of load. System protection and relaying are intended to achieve these objectives.

10.2 Changing automatic generation control objectives

In the new era of power industry restructuring and deregulation, some of the methods for automatic generation control will have to be revised. One new requirement is to accomplish the additional task of implementing open access regulation contracts. Currently, electric utility companies, market participants, the staff of the North American Electric Reliability Council's (NERC) regional councils, and others are considering various approaches for accomplishing regulation and allocating the associated costs.

Traditionally, most power transaction agreements included the costs for regulation and other ancillary services (or interconnected operating services) as part of the price for power supply services. Order 888 issued by the Federal Energy Regulatory Commission on April 24, 1996 created requirements to separately identify seven ancillary services, including the costs for regulation and frequency response services.

A number of factors need to be taken into account to determine the cost of regulation and frequency response services, including:

1. The cost of operating generating units that provide regulating services
2. Costs for reserve transmission capacity to accomplish regulation

3. Costs for control and telemetry equipment needed to accomplish regulation
4. Costs for wheeling, when appropriate

Models have been proposed for calculating regulation service requirements using statistical data. Since regulation cannot be achieved for all possible load variations, such models require an assumption about the percentage of the time for which regulation is to be accomplished.

10.3 Control performance criteria

NERC has recognized the need to improve its control performance criteria. The old NERC "a1" criteria required that a control area's ACE return to zero every 10 minutes. The 10-minute period coincided with the definition of "operating reserves," that is, reserves used to make up power during contingency events. The old NERC "a2" criteria had to do with the average value of ACE between zero crossings. Averaging ACE was intended to avoid having generation follow very short-term load swings or noisy signals in instrumentation systems. Criteria a1 and a2 were based on operating objectives and helped in system control, but these criteria did not distinguish between ACE values causing increased frequency deviations and ACE values helping to return frequency to scheduled values. More importantly, NERC recognized that criteria a1 and a2 involved economic costs, which are difficult to justify in a competitive business environment.

The new NERC control performance standards are "CPS1" and "CPS2." CPS1 uses an index calculated by taking the average value of ACE, divided by 10, multiplied by the frequency bias constant, multiplied by the difference between actual system frequency and scheduled system frequency. CPS1 recognizes that control areas should increase generation during periods when overall system loads are increasing and should decrease generation during periods when overall system loads are decreasing. CPS2 was designed to allow larger control areas to have greater ACE deviations than smaller control areas during each 10-minute period. Thus, CPS2 recognizes that control areas function as part of the interconnected system and share the responsibility for regulation.

NERC has not only changed the technical definitions for control performance, but also has established procedures to better monitor the performance of individual systems and to make individual systems comply. NERC's vision is to become an independent industry self-regulatory organization that will enforce compliance with reliability standards in a fair and nondiscriminatory manner. At the conclusion of this transition, NERC will be renamed the North American Electric Reliability Organization (NAERO).

Deregulation, restructuring, and competition require modifying the original formulation of the automatic generation control problem and the associated methods of evaluating control performance. Electric power industry restructuring will result in new definitions for control areas and will require innovative methods for calculating and allocating the costs for regulation and frequency response services. These tasks are being addressed by NERC committees and by the new participants in power markets.

10.4 End-of-chapter questions

1. What is the difference between the terms "Automatic Generation Control" and "Load Frequency Control?"
2. How many control areas are there in North America, and what is the function of a control area?
3. How does Automatic Generation Control maintain system frequency near the scheduled value (usually 60 Hz) during normal conditions? How is frequency regulated during emergency or disturbance conditions?
4. What effect will FERC Order 888 have on the way generation is controlled and costs are allocated in the electric power industry?
5. Why did the North American Electric Reliability Council change the methodology for calculating Control Performance Criteria?
6. Can you suggest an alternative formulation for control performance criteria? What are the strengths and weaknesses of the approach you are suggesting as compared to the NERC criteria?

chapter eleven

New reliability and control concepts

11.1 The layman's definition of reliability

Webster's Dictionary defines reliability as "the quality or state of being reliable."

The word reliable is, in turn, defined as "suitable or fit to be relied on." In defining the word rely, *Webster's* refers to trust, confidence, and dependability. So, in other words, power systems are reliable if they can be trusted, if we have confidence in their performance, and/or if we can depend on power being available when we throw the wall switch.

As long as electric power service is reliable, most people are content and unconcerned about pursuing the subject of reliability. However, the general level of interest in reliability peaks when electric power service is *unreliable*. Consequently, our vocabulary for describing unreliability is somewhat richer than our vocabulary for describing reliability.

Unreliable conditions are referred to as "service interruptions," "power disruptions," "power outages," "power failures," "blackouts," and "brownouts." All of these terms were used by those who feared that power industry deregulation would have an adverse effect on power system performance.

11.2 The academic and traditional definitions of reliability

Textbooks on power system reliability often begin with a classical definition, such as "The reliability of a power system is the probability that the system will perform its intended function in an acceptable manner, for some intended period of time, under specified operating conditions."

Within the field of power system reliability analysis there are well-established methods for analyzing and calculating component and system reliability. At the bulk electric system level, U.S. transmission grids have been more than 99.9% reliable. This can be interpreted to mean that the systems are unavailable less than 2 hours per year.

At the power distribution system level, a number of indices have been defined for assessing reliability. The system average interruption duration index, or SAIDI, provides a measure of the average length of time that customers are without power. SAIDI is normally expressed in minutes per year. The system average interruption frequency index (SAIFI) is closely related to SAIDI. SAIFI provides a measure of the number of times when customers lose power. SAIFI generally is expressed in terms of the average number of outages per year. The customer average interruption duration index (CAIDI) is still another index that may be calculated by dividing SAIDI by SAIFI. Alternatively, CAIDI may be directly calculated as the average length of time when customers are without power each time there is a power supply interruption. CAIDI is usually expressed in minutes.

11.3 North American Electric Reliability Council reliability definitions

The North American Electric Reliability Council (NERC) used the following definitions of power system reliability and power system security during the 1980s and 1990s. NERC said that reliability has two components: adequacy and security. Adequacy has to do with the question, "Will there be enough generation and transmission to meet load levels and customer requirements?" Security has to do with the question, "Will the generation and transmission systems operate reliably in the sense that they perform their intended functions?"

11.4 Traditional power system operations in control areas

Power system control has traditionally been accomplished by entities in "control areas." The "control area" concept involves balancing load with generation (taking into account schedules for interchanges with neighbors) within the electrical boundaries corresponding to the service territory of a vertically integrated utility. Today there are approximately 150 control areas within the North American bulk electric system. NERC defines a control area as "An electrical system bounded by interconnection (tie line) metering and telemetry. It controls generation directly to maintain its interchange schedule with other control areas and contributes to frequency regulation of the interconnection."

When the amount of generation in a control area (including net imported power) exceeds the total load within a control area, frequency will increase. When the amount of generation (including net imported power) is less than the total load within a control area, frequency will decrease. Small variations in frequency can be tolerated, but large frequency variations cause protective relays to operate, causing customers to experience service interruptions or "blackouts." The continuing steady state balancing of load and generation is therefore essential to system reliability.

The balancing of load and generation must be performed on an instantaneous basis aided by automatic generation control systems, as well as effective planning in the operations planning time frame (minutes to months) and effective system planning in a time frame of years to decades. The quality of load forecasts is critical to this balancing function. A knowledge of generation maintenance schedules, forced outage rates, and power purchase contracts is also critical to this function.

11.5 The new paradigm: operating and service functions

The control area approach has worked well as a means of maintaining reliability and meeting the economic needs of electric utility customers (native load customers), as well as a basis for managing wholesale purchases and sales to utilities and using the transmission system for "wheeling services." However, with the advent of deregulation and restructuring, the control area approach has proven to be inadequate in several respects. First, the large number of control areas (approximately 150) made it difficult to coordinate security operations. Recognizing this, NERC created a hierarchical system and established 22 security coordinators, each with coordinating responsibility for large geographic areas, usually involving multiple control areas.

The traditional control area approach also was inadequate, because it was based on the vertically integrated model. It did not take into account the new structure involving unbundled systems with generation, transmission, and customer service functions separated with a utility or offered by entities not under a single corporate ownership umbrella. Additionally, the control area approach needed to be redefined to accommodate innovative approaches taken by power marketers, such as having a single generating unit with power contracts to serve loads in other control areas constitute a control area.

Recognizing a pressing need to rethink control area concepts and define "functional responsibilities" rather than "organizational responsibilities," NERC formed a Control Area Criteria Task Force that produced a final report in 2001. This report identified nearly 100 operating functions that needed to be performed by some entity to maintain reliability and accommodate the new open-access, competitive power markets. One of the reasons for doing this was to be able to assign the various functions to existing and new or emerging organizations, and thereby clarify the responsibility for maintaining system reliability. Of course, it was impossible to anticipate how future organizations may be structured, but the NERC task force believed that the following list of entities currently performing operating functions was reasonably complete as of the date of the final report in 2001:

1. Generators*
2. Transmission service providers
3. Transmission owners*
4. Transmission operators*
5. Distribution providers

6. Load-serving entities*
7. Purchasing/selling entities*
8. Security authorities
9. Balancing authorities*
10. Interchange authorities*
11. Compliance monitors

Asterisks have been added to the above list to indicate the functional roles usually considered to be associated with "control areas." In other words, there are four functional roles without asterisks (transmission service provider, distribution provider, security authority, and compliance monitor). These four functional roles have assumed importance in the new deregulated and restructured electric power industry. The NERC task force report also separated "service functions" from "operating functions." With this bifurcation, of the four functions without asterisks, the transmission service provider function, the security authority function, and compliance monitor function are "service functions," while the distribution provider function is an "operating function."

11.6 End-of-chapter questions

1. What are three terms commonly used to mean that an electric power system has been unreliable?
2. If a 500-kV transmission system has been 99.8% reliable during a given year, how many minutes of unavailability have been experienced?
3. In designing a backup, on-site power supply facility for a hospital, an engineer would like to maintain enough fuel on site for usual outage conditions with a 50% safety factor. Would the engineer need data about SAIDI, SAIFI, or CAIDI?
4. What is meant by the term "power system adequacy," as defined by NERC? Do you think this term is more related to system planning or system operations?
5. What is meant by the term "power system security," as defined by NERC? Do you think this term is more related to system planning or system operations?
6. What conditions will cause frequency to increase within a control area? What could be done to arrest or reverse frequency increases within a control area?
7. How many control areas are there currently in North America? Does it seem practical to coordinate with this number of entities in making a real-time decision to be implemented with a 10- or 15-minute time window?
8. Explain why NERC has defined "operating functions" and "service functions" as a better means of assigning responsibilities after the break up of the vertically integrated traditional electric utility system structure.

chapter twelve

Available transfer capability

12.1 A new methodology for assessing transmission line limitations

When the interconnections in the electric utility industry began to experience increased power flows as a consequence of industry restructuring, it became apparent that improved methods would be needed to calculate "available transmission transfer capabilities" (sometimes called available transfer capability, or ATC).

There are three limiting factors to be considered in determining transmission transfer capability. First, "thermal limits" must be taken into account. Thermal limits have to do with the amount of electrical current that a transmission line or flowgate can accommodate. Of course, the magnitude of current in a transmission line or flowgate can be determined by dividing the total amount of electric power transfer by the voltage level of the transmission line or flowgate. Thermal limits are determined for a specified time period. In other words, a transmission line or flowgate can accommodate a specific amount of current for a certain period of time before overheating.

"Voltage limits" are the second limiting factor. For every transmission line or flowgate there are minimum and maximum acceptable voltage limits. If voltage goes too high or too low, electrical power equipment may be damaged or protective-relaying systems may open breakers, causing customer outages or even widespread cascading blackouts. Most of the textbooks used for the first course in power systems analysis explain why system voltage variations are more sensitive to reactive power flows than to active power flows. These textbooks also typically discuss some of the techniques for maintaining acceptable voltages and var flows, including the use of synchronous machines, static capacitors, tap changing transformers, etc.

"Stability limits" are the third limiting factor. Following a power system fault, very high current flows may occur during the subtransient period (generally considered to include approximately the first 0.05 sec or the first 3 cycles for a 60-Hz system) or dynamic period (the period after the subtransient period, but prior to returning to the steady state, generally considered in the range of milliseconds up to a few minutes). Transmission elements or flowgates must be capable of withstanding the initial fault currents, as well as effects of associated subsequent frequency and voltage oscillations. Stability limits take these factors into account and also consider whether or not the system can establish a new, acceptable, steady-state operating point.

Consideration of the effects of a contingency involves examining thermal, voltage, and stability limits and the way these limits shift as a function of different operating conditions. The determination of ATC can at best be based on an approximation of these effects and the use of judgement to identify the most likely contingencies and system variations over time.

Recognizing the importance and complexity of these issues, the North American Electric Reliability Council (NERC) took the lead in developing uniform ATC definitions and ATC methodologies. NERC'S ATC definitions and methodologies are important because they provide a common framework to be used by those industry participants who are primarily interested in maintaining bulk electric system reliability, as well as those industry participants who are primarily interested in the commercial viability of wholesale markets. The NERC ATC definitions and methodologies also responded to Federal Energy Regulatory Commission (FERC) concerns expressed in FERC Notices, Orders, and staff reports. FERC Notices require that information about ATC be made available to all transmission users. The NERC term "Available Transfer Capability" addresses FERC requirements, but has a more detailed and complete definition and application.

NERC defined ATC as "a measure to the transfer capability remaining in the physical transmission network for further commercial activity over and above already committed uses."

An alternate definition of ATC is the total transfer capability minus three factors: the transmission reliability margin, the capacity benefit margin, and the sum of existing transmission commitments.

Transmission transfer capability (TTC) is defined by NERC as the amount of electric power that can be transferred over the interconnected transmission network in a reliable manner while meeting all system conditions (before and after likely contingencies).

Transmission reliability margin (TRM) is defined by NERC as the amount of transmission transfer capability necessary to ensure that the interconnected transmission network is secure under a reasonable range of uncertainties in system conditions.

Capacity benefit margin (CBM) is defined by NERC as the amount of transmission transfer capability reserved by load-serving entities to ensure access to generation from interconnected systems to meet generation reliability requirements.

TTC, TRM, CBM, and ATC are typically determined based on experience, knowledge of likely contingencies, expectations for transmission commitments, and computer simulations of power system operating conditions. Since the 1950s, simulation studies have been used to prepare system operators for likely operating conditions. Modern simulation studies are much more accurate and sophisticated than the earlier simulation studies, using more telemetered data, more powerful computers, and improved power system simulation software. Having a number of simulation studies available provides all market participants with prior knowledge of the characteristics and capabilities of the power system.

12.2 Guiding principles for ATC calculations

When NERC issued its report on "Available Transfer Capability Definitions and Determination" in June 1996, it was noted that individual systems, power pools, subregions, and regions would be permitted to develop their own procedures for determining or coordinating ATCs based on a regional or wide-area approach, as long as these procedures were consistent with the following six principles in the NERC ATC report They have been reworded and abbreviated from the NERC report to facilitate presentation to university students:

1. ATCs should realistically indicate the actual transfer capabilities available to the electric power market. ATCs must be accurate and realistic to provide a basis for market decisions, particularly in areas where there is significant congestion or where there are many wholesale power purchases and sales.
2. ATCs should recognize that power flow conditions vary in time and are affected by "simultaneous transfers" and "parallel path flows" on the interconnected transmission network. The ATC of a collection of lines will generally be less than the ATC found by adding the ATCs of the individual lines.
3. ATCs should take into account the direction of power flows on transmission lines and whether active power is injected or extracted at generation and load busses. The ATC from point A to point B is not necessarily equal to the ATC from point B to point A.
4. ATC calculations and results should be coordinated and openly shared on a regional basis.
5. ATC calculations should be consistent with NERC, regional, subregional, power pool, and individual system planning and operating policies.
6. ATC calculations should take into account uncertainties in system conditions and provide operating flexibility.

12.3 End-of-chapter questions

1. Under what conditions, or in what kind of systems, is the accurate calculation of available transmission system transfer capability most important?
2. How does available transfer capability (ATC) relate to total transfer capability (TTC)?

3. What is the purpose of power system simulation studies, and how are these studies used in determining ATC?
4. Why are accurate ATCs important to those who are primarily concerned with maintaining reliability? Who are some of the entities with these concerns?
5. Why are accurate ATCs important to those who are primarily concerned with commercial transactions in electric power markets? Who are some of the entities with these concerns?
6. When should "time-variant power flow conditions" be taken into account in calculating ATCs? Make up an example, including variable megawatt flows on a flowgate, to explain your answer.
7. Why should "simultaneous transfers" be taken into account in calculating ATCs? Make up an example, including a network configuration and megawatt flows on flowgates experiencing "simultaneous transfers" to explain your answer.
8. Why should "parallel path flows" be taken into account in calculating ATCs? Make up an example, including a network configuration and variable megawatt flows on the flowgates, to explain your answer. Show how the transfer capability of a single transmission line can be affected by flows on other lines that are part of the interconnected network.
9. Can the ATCs of individual transmission lines be added to determine the ATC of an interface between two systems? Why not? Would the aggregated ATC be generally greater than or less than the sum of the ATCs?
10. Why should "uncertainties in system conditions" be taken into account in calculating ATCs? Make up an example, including variable megawatt flows on a flowgate, to explain your answer.

chapter thirteen

Network congestion and transmission loading relief

13.1 The network congestion problem

Shortly after it became apparent that the electric utility industry would be restructured to create increased competition and provide open access to transmission, it was recognized that increased network congestion could be one of the primary problems associated with electric industry restructuring.

It was anticipated that industry restructuring would lead to a greatly increased number of transactions to purchase and sell electricity on the grid. As a consequence, electric power systems would experience more frequent transmission line overloads. This has certainly proven to be the case, as discussed at the end of this chapter.

Early in the 1990s, the North American Electric Reliability Council (NERC) and the Electric Power Research Institute (EPRI) organized industry efforts to consider various approaches for reducing network congestion in a way that would be equitable and fair to all of the parties involved.

Network congestion can be reduced by cancelling transactions, redispatching generation, reconfiguring transmission, or reducing loads. Obviously, there can be very great financial impacts from taking any of these actions. And, of course, equal or even greater financial impacts can result from allowing network congestion to cause overloads or other operating security limit violations. Consequently, all participants in electric power markets have taken a keen interest in network congestion problems and methods for relieving network congestion.

13.2 The transmission loading relief approach

Following extensive debate within the NERC committee structure, NERC adopted transmission loading relief (TLR) procedures as the primary means to be used by security coordinators for addressing network congestion problems. The TLR approach is intended as a means to mitigate potential and/or actual violations of operating security limits while honoring transmission service reservation priorities. For the purposes of this discussion, operating security limits are defined by NERC in the current version of the NERC online operating manual as "The value of a system operating parameter (e.g., total power transfer across an interface) that satisfies the most limiting prescribed pre- and post-contingency operating criteria as determined by equipment loading capability and acceptable stability and voltage conditions."

NERC's TLR procedure follows the Federal Energy Regulatory Commission's "Pro Forma Tariff." The purpose of the NERC TLR procedure is to define the actions and communications required to safely and effectively reduce the flow on a flowgate. In this discussion, a flowgate is defined as a transmission element of the bulk electric system. NERC has defined the following TLR levels as a basis for the TLR procedure:

1. TLR Levels 1: Notification
2. TLR Level 2: Hold
3. TLR Levels 3a and 3b: Curtailment using nonfirm
4. TLR Level 4: Reconfiguration
5. TLR Levels 5a and 5b: Curtailment using firm
6. TLR Level 6: Emergency procedures
7. TLR Level 0: TLR concluded

A security coordinator calls a TLR Level 1 when the current loading trends are likely to result in an operating security limit (OSL) violation. In the case of a TLR Level 1, other security coordinators, transmission providers, control areas, and merchants are notified, but there is no substative effect on interchange transactions.

A security coordinator calls a TLR Level 2 when a flowgate is at or approaching an operating security limit violation, and a transmission provider receives a request to implement new interchange transactions that will cause an overload. A TLR Level 2 may also be called when a security coordinator anticipates problems and requires time to analyze possible solutions. The effect of a TLR Level 2 is to cause all schedules that impact identified flowgates to be held at their current active power flow levels. In other words, actions are taken so that the megawatt (MW) loading of flowgates is not increased.

A security coordinator calls a TLR Level 3 when a flowgate is at or approaching an operating security limit violation, and a transmission provider receives a request to implement new interchange transactions and/or increased interchange transactions that will cause an overload. The effect of a TLR Level 3 is to curtail interchange transactions using nonfirm, point-to-point transmission service and allow higher priority transactions to start or increase.

It should be noted that the curtailment of interchange transactions using firm, point-to-point transmission service involves having security coordinators coordinate with transmission providers to identify the redispatch options that a transmission customer could use to reduce the loading on the particular flowgates of interest. If the options identified do not sufficiently reduce loading, calculations are made to determine the MW curtailment that can be achieved using the TLR procedure.

A security coordinator calls a TLR Level 3a to curtail lower-priority nonfirm, point-to-point transmission service and to allow higher-priority transactions to start or increase. A security coordinator calls a TLR Level 3b to curtail firm service and mitigate an operating security violation.

A TLR Level 4 is called when a flowgate is above its operating security limit and there are no nonfirm interchange transactions that can be curtailed to solve the problem. The effect of a TLR Level 4 is to reconfigure transmission systems and avoid curtailing firm interchange transactions.

A security coordinator calls a TLR Level 5 when a flowgate is at or approaching an operating security limit violation, and a transmission provider receives a request to implement new interchange transactions and/or increased interchange transactions that will cause an overload. The effect of a TLR Level 5 is to curtail interchange transactions using nonfirm, point-to-point transmission service and allow higher-priority transactions to start or increase.

A security coordinator calls a TLR Level 5a to curtail lower-priority, nonfirm, point-to-point transmission service and to allow higher-priority transactions to start or increase. A security coordinator calls a TLR Level 5b to curtail firm transmission service and mitigate an operating security violation.

A security coordinator calls a TLR Level 6 when the actions associated with TLR Levels 3, 4, and 5 are insufficient to resolve the problem or when the flowgate reaches such a critical level that emergency actions are required.

The effect of a TLR Level 6 is to direct control areas or transmission providers to take actions, such as generation redispatch, transmission reconfiguration, or load shedding, to either mitigate the critical condition or provide time for the actions associated with TLR Levels 3, 4, and 5 to take effect.

TLR Level 0 indicates that the loading on the flowgate is continuing to trend downward and that any operating security limit violations have been addressed. The purpose of TLR Level 0 is to provide a means to notify other security coordinators, transmission providers, control areas, and merchants that all curtailed transactions can be restored.

It should be noted that during any of the TLR levels, transmission providers are not obligated to redispatch their own resources to maintain transactions using firm, point-to-point transmission service prior to being curtailed, according to the FERC Pro Forma Tariff.

13.3 Criticisms of the TLR approach

Market participants have complained that the TLR approach has several deficiencies. The complaints primarily identify cases in which there are unnecessary curtailments, a lack of standardized protocols for providing information, and/or discriminatory conduct.

FERC has produced staff reports documenting these complaints and suggesting the need for modifications in the TLR procedures. FERC staff reports indicate that the majority of complaints focus on the possibility that the TLR procedures are not providing open access to transmission systems in the Midwest Region. The Midwest Region is likely to have network problems because there are large power flows over long distances in this region. In addition, the coordination of remedial actions is complicated by the fact that there are many control areas (61 in the year 2000) and security coordinators (6 of 22) in this region.

The transmission systems in the Midwest were designed and built to:

1. Allow vertically integrated utilities to serve their own native loads
2. Provide a means for intercompany sharing of capacity reserves during emergencies
3. Reduce power generation investments

The transmission systems in the Midwest were *not* designed and built to:

1. Transfer large amounts of power over long distances
2. Accommodate power marketing in the deregulated electric utility industry as directed in FERC Orders 888, 889, and 2000

NERC recognizes that the TLR process is not the ideal solution to the network congestion problem, but it is being used as an interim solution, and it is being continuously modified as problems are identified.

13.4 Network congestion data

Current data concerning the increases in wholesale power receipts (purchased power plus exchanges received and wheeling received) is available from the U.S. Department of Energy's Energy Information Administration at the following Web site: http://www.eia.doe.gov/cneaf/electricity/.

Department of Energy data show dramatic increases in power purchases and sales in recent years following electric industry restructuring.

13.5 End-of-chapter questions

1. Why has electric power industry restructuring caused increased network congestion?
2. What steps can be taken to reduce network congestion?
3. Which industry organization developed the Transmission Loading Relief (TLR) procedures used by the industry?
4. According to NERC, are both precontingency and postcontingency operating criteria considered in determining if operating security limits have been violated?

5. According to NERC, are system stability conditions as well as steady-state performance used in determining if operating security limits have been violated?
6. What actions are taken when a security coordinator calls a TLR Level 1?
7. What actions are taken when a security coordinator calls a TLR Level 2?
8. What actions are taken when a security coordinator calls a TLR Level 3?
9. What actions are taken when a security coordinator calls a TLR Level 5?

chapter fourteen

The use of power flow and stability analysis tools

14.1 Operating security limit (OSL) violations

Power system operators maintain the reliability of their systems by anticipating and/or correcting operating security limit violations. An operating security limit violation will not necessarily jeopardize the reliability of the Interconnection or create a widespread problem or cascading blackout, but the existence of an operating security limit implies one of three things:

1. A steady-state rating of a monitored element has been exceeded
2. A voltage limit has been exceeded
3. A stability limit has been exceeded

When any one of these conditions is identified, a component or system failure is imminent. If the failure has not yet occurred, then it is likely that it will occur in the future with additional loading on the power system or with the added stress of a contingency occurring.

14.2 Tools for determining OSL violations

One question related to this discussion is, "How are the limiting conditions established?" Information about the steady-state limiting conditions can be obtained from a power flow analysis. Power flow analyses require information about system configuration (usually in the form of a bus admittance matrix or a bus impedance matrix), information about the net active and reactive power injections at each bus, and information about regulating transformers. The result of a power flow analysis (or the output of a power flow program) is the active and reactive power flow in each flowgate and the voltage magnitude and angle at every modeled bus. System operators anticipate the steady-state effects of increased power system loading or the occurrence of a contingency using power flow analysis tools.

Information about stability limits is obtained from stability analyses. Most of the modern undergraduate power system analysis textbooks devote a chapter to stability analyses. In addition, many other textbooks have been written exclusively about this subject. Stability analyses fundamentally begin with a "swing equation" representing the behavior of a generating unit following a fault or disturbance. Usually, relative stability can be determined from calculating how the machine's rotor angle changes during a very short period of time following a fault or disturbance. Normally, some assumption is made about fault-clearing time, and the key parameter becomes the angular change prior to fault clearing.

For systems with more than two machines, hand calculation becomes impractical, but today many computer programs are available for performing stability analyses. Usually, a power system analysis software package will include a power flow program, a fault analysis (or short-circuit analysis program), and a stability analysis program. Modern versions of these programs are very user-friendly. Models for most equipment are found in the libraries provided with the programs, and the models are modified using manufacturer data.

14.3 End-of-chapter questions

1. Will exceeding an operating security limit violation necessarily jeopardize the reliability of the Interconnection or create widespread problems or cascading blackouts? Explain.
2. Operating security limits are exceeded under three conditions. What are the three conditions?
3. How do power system operators anticipate the steady-state effects of increased power system loading or the occurrence of a contingency?
4. Why is fault-clearing time taken into account in analyzing the stability conditions in a power system?
5. Do you think you could determine the stability of a power system with six machines during a 1-hour examination? Explain if or when this may be possible.

chapter fifteen

Technology needs for the electric power industry

15.1 Opportunities and threats

Deregulating and restructuring the North American electric power industry have been advocated as a means of introducing new technologies into the industry. Advocates say, "Look at the effect of deregulation in the telephone industry." It is a fact that technological changes did occur in the telephone industry on the heels of deregulation. Similarly, it seems fair to say that deregulation and restructuring will create opportunities and incentives to introduce new technologies in the North American electric power industry.

Deregulation and restructuring also pose threats in the sense that the reliability and cost of electric service may be adversely impacted if more sophisticated new technologies are not available when they are needed to address the additional problems and complexities associated with deregulation and restructuring. These concerns seem well justified after the serious problems in California in 2000 and the Western system power outages of July and August 1996. Many articles have addressed the potentially adverse reliability implications of restructuring. Clearly, the new business environment will create not only new forms of regulation and competition, but also unprecedented needs for technological solutions to problems in the areas of systems operations and systems planning.

15.2 Lessons from the past

Following the Great Northeast Blackout of 1965, the Federal Power Commission identified several needs for new technologies and techniques. The Federal Power Commission said the industry needed better regional coordination, more transmission where large load centers were separated from generation facilities, better load forecasting techniques, and computer simulation tools for system operators. Many of these technologies and techniques were subsequently addressed by the industry to maintain and enhance the reliability of the power systems during that era.

With deregulation and restructuring, the business environment has changed dramatically. Current systems have little resemblance to the systems that existed in 1965 when the Northeast Blackout occurred. Deregulation and restructuring have created new challenges because power systems have to operate in a way that was never envisioned when they were planned, designed, and constructed. To make a rough analogy, this is like taking a Toyota Camry designed for highway use and trying to use it as an off-road, rough-terrain vehicle. With the new, profoundly changed operating environment, it is interesting that many of the technology needs identified by the Federal Power Commission following the Northeast Blackout are now again key needs as a consequence of deregulation and restructuring.

15.3 An overview of the problem

Requirements for open access to transmission networks have increased power flows and changed power-flow patterns. To cope with the increased and changed flows, innovative means are needed for upgrading transmission networks and for using existing networks to their fullest potential. Several individuals and organizations, including the Electric Power Research Institute (EPRI), have been placing a significant amount of effort on the development of methods and hardware systems using power electronics to address these needs. Notable in this area is the development of flexible A.C. transmission systems (FACTS).

Unbundling and increased competition require new information systems to be developed both for economic power-exchange-related functions and for system-security-related and reliability coordination functions. Improved transducers, data acquisition systems, and expanded or enhanced communications technologies will be needed.

New technical training requirements and engineering analysis tools will also be needed to deal with other system operations challenges posed by restructuring. Many of the committees within the North American Electric Reliability Council's (NERC) committee structure have been addressing the increased complexity and dimensionality of problems in a restructured industry. These efforts will continue into the next decade.

Recently, a number of mergers and acquisitions have involved electric power companies and natural gas companies. At a time when it is critically important to be able to assign operational responsibilities and authorities, it is becoming increasingly difficult to keep up with industry structural changes. Reorganizational trends are likely to continue and accelerate in the future. Currently, about 200 investor-owned electric utility companies exist in the United States. Industry analysts have suggested that there may be as few as 50 investor-owned electric utility companies by the end of the next decade. Examples of the trend toward corporate combinations abound. Electric power companies are acquiring natural gas distribution companies and forming new integrated energy companies. Conversely, large natural gas companies are buying electric utility companies.

The time frame for restructuring is a very critical issue from the electric utility point of view, because more time can allow major changes in utility operations. The states with the highest rates have generally shown the greatest interest and activity to initiate electric industry restructuring. In most cases, the states with the highest rates currently have the most difficult operating problems.

The Federal Energy Regulatory Commission (FERC) orders and rulemakings have clarified several restructuring issues, but have also introduced new issues, such as:

1. Defining what is meant by comparable open transmission access, particularly when power pools are involved
2. Functional unbundling to separate transmission system operators from wholesale marketers
3. Allocation of the costs for the real-time information networks
4. Methods for maintaining system reliability, with the new demands being placed on the transmission grid

NERC and the Electric Power Research Institute (EPRI) have focused on the adequacy of transmission capacity and transfer capability determinations, and associated technical problems and challenges. Both organizations have also investigated the various aspects of the recent power outages to determine if restructuring trends are root causes.

One of the technical concerns about power system reliability in the future has to do with the fact that generation capacity reserve margins are declining. Generation reserves are declining for several reasons. The power plants built in the middle 1960s or earlier are reaching the end of their useful lives. Some of these plants are being retired, and others have been derated. Life extension and uprating efforts are being undertaken in some cases, but the net effect is reduced available capacity. The amendments to the Clean Air Act and other environmental requirements are resulting in plant retirements or the nonuse of coal-fired units. Some of these coal-fired units are 500 megawatts (MW) or larger. Hence, a significant amount of generating capacity is unavailable for environmental reasons. Industry restructuring and the trend toward increased competition are forcing the retirement of units that are not competitive from an operating cost point of view. There is uncertainty about the treatment of stranded assets, but unquestionably some of the stranded-generating assets will be removed from service. Electric industry restructuring has stimulated the construction of nonutility generation or exempt wholesale generation (EWG), but has reduced the construction of new utility-owned generating capacity. During the period of transition to increased regional coordination, capacity may not be built where it is needed and when it is needed.

Transmission capacity reserve margins are also declining. Restructuring is increasing the number of transactions and the amount of power purchased and sold over interconnections. Transmission systems were designed to accommodate the amount of wheeling associated with the former system of regulation (vertically integrated utilities serving franchise areas without open access). New transmission lines face a great deal of environmental opposition. Not only are individuals and groups concerned about aesthetics and electrical safety, but, increasingly, there is a concern about adverse human health effects from electric and magnetic fields (EMF). The EPRI and other research organizations have found ways to reduce transmission line fields, improve aesthetics, and address safety considerations. However, the costs associated with these methods are significant. Uncertainties about industry restructuring and the trend toward increased competition discourage the construction of new transmission lines. Transmission is a very expensive, high-risk investment. As in the case of generating capacity, during the period of transition to increased regional coordination, transmission capacity may not be built where it is needed and when it is needed.

While relatively minor improvements are being made in the equipment for generating and transmitting electricity, major improvements are being made in the information technologies used to monitor and control market transactions and power flows. The hardware and software used for data acquisition have become much more sophisticated and much less expensive. Electric utilities are installing fiber optic communication systems with new transmission construction along existing transmission corridors. Fiber optic methods complement existing extensive communication systems comprised of power line carrier, microwave communication, and radio/wireless systems. Research efforts have focused on improved database management and open systems approaches to reduce the costs of telecommunications functions. The cost of computers and data processing power has reduced exponentially in the last decade. This cost reduction, along with advances in operating systems and improved application software, provides a powerful platform for the power system operation tools and analytical capabilities needed to address the transition problems associated with restructuring. Information technologies will make it possible to develop more cost-efficient and environmentally effective methods for power generation by reducing the losses associated with the transmission of electricity, and by designing power systems that accommodate increased economy interchange and facilitate environmental dispatching. Updated baseline information and forecasts for the next decade on the environmental impacts and costs and availability of conventional and unconventional energy production systems are needed. Particular emphasis should be placed on opportunities for the increased use of combined cycle natural gas generation, wind energy, solar thermal energy, and photovoltaic energy.

Current public policies and possible policy initiatives should be examined for all of the power-generation and energy-storage technologies. As the electric utility industry transitions toward a market-driven, competitive approach to power generation, there is a possibility that long-term cost effectiveness may be sacrificed for short-term economies. In addition, there are indications that some of the environmental programs that have been initiated by state public service commissions may not be continued with a restructured regulatory approach.

There is a need to develop improved methods for automatic generation control in a restructured industry and begin the development of more specific proposals and research plans for redesigning each of the other existing energy control center applications.

Unquestionably, with increased "wheeling," the losses associated with the transmission of electricity may become more significant. However, new technologies are currently being developed to reduce losses in individual lines and to give power system operators increased control over power routing. Development of better data and forecasts concerning these technologies as a basis for influencing future technical and policy decisions is needed. Several specific design concepts are now being considered, including the cost and value of replacing mechanical switching systems with new power semiconductors, a technology called FACTS flexible transmission systems. Consideration is also being given to using new materials and new configurations for transmission line design. In addition, several reports have been published on the potential for upgrading existing transmission lines.

The unbundling of power systems services and the transition to having an independent system operator will create opportunities to redesign power system operating procedures to accommodate increased economy interchanges and facilitate environmental dispatching. There is a need to compare the informational requirements and operational guidelines currently used in utility power system control centers with the requirements and guidelines that will best serve the public interest when independent system operators assume the responsibility for economic dispatching decisions. It will be important that redesigned systems and procedures properly incorporate the need for maintaining power system reliability and power quality.

As restructuring is implemented, each of the existing energy-control center applications will need to be revised, and in some cases totally redefined, for the restructured industry to function effectively. Existing energy-control center applications include economic and environmental dispatching, automatic generation control, unit commitment, interchange evaluation, power flow, contingency analysis, state estimation, supervisory control, and data acquisition.

The opportunities associated with open access have to do with the various options for reducing energy costs or preventing the future increase of electricity rates. The options are summarized here:

1. Negotiate favorable long-term contracts with existing energy suppliers during the period when new legislation is being considered. (This has been the strategy of one of the big three automakers in its Michigan operations.)

2. Promote legislation that encourages multiple energy suppliers to compete for industrial customers, and then select the lowest-cost offerings. (This is the strategy advocated by the Electric Power Supply Association and others.)
3. Purchase or construct electricity-generation facilities and become a self-generator for some or all of your own needs, or even become an electric wholesale generator, selling power into the grid for profit.

The threats associated with the new business environment involve the possibility of reduced system reliability or increased energy costs as a consequence of strategic decisions. The options available to prevent or mitigate these threats are:

1. Assure that reliability is adequately addressed in new contracts for energy supply
2. Promote legislation and regulatory reforms that address reliability, supply adequacy, and power-system security
3. Recognize that a new system of marketing or self-generation options will not necessarily reduce energy costs or provide sufficient levels of reliability

15.4 Summary

Obviously, many technological concerns are associated with restructuring that has not been addressed here. Some of the existing programs in support of environmentally preferred technologies have been discontinued. Most troubling are concerns that broader research and development will not be supported.

To conclude with an understatement, the next few years will certainly be interesting. The forces of change have been unleashed, and the effects will undoubtedly be profound. It will be important that all electricity consumers stay apprised of new developments in the area of electric power industry and carefully evaluate strategies for coping with restructuring.

15.5 End-of-chapter questions

1. Why are the new information technologies essential for accomplishing the goals of open access, competition, deregulation, and restructuring?
2. Do think that deregulation/restructuring will increase the use of new technologies in the power industry? Explain.
3. Can you think of a situation in which a piece of equipment or a system was designed for one set of conditions, but then had to be used for some other set of conditions?

4. Some people are concerned that competition will result in less information sharing among power system engineers. What can be done to assure that power engineers receive adequate training after deregulation/restructuring?
5. Some people are concerned that competition will result in reduced expenditures for research and development at a time when new technological solutions to unprecedented problems are critically needed. What can be done to assure that research and development efforts are sufficiently funded?

Index

F

G

H

I

J

K

L

M

N

O

P

Q

R

S

T

U

V

W

X